NATIONAL GEOGRAPHIC
SCIENCE

SCIENCE INQUIRY
AND WRITING BOOK

SCIENCE

NATIONAL
GEOGRAPHIC

School Publishing

PROGRAM AUTHORS

Judith S. Lederman, Ph.D.

Randy Bell, Ph.D.

Malcolm B. Butler, Ph.D.

Kathy Cabe Trundle, Ph.D.

David W. Moore, Ph.D.

Science Inquiry

Life Science

Life Science

Science Inquiry

Life Science

Earth Science

Science Inquiry

Earth Science

Earth Science

Science Inquiry

Physical Science

Physical Science

Science Inquiry

Physical Science

Physical Science

Science in a Snap!

Science in a Snap! Sorting Seeds

Observe a variety of seeds. Sort seeds that appear to be from the same kind of plant into groups. **Compare** the seeds in each group. How do they differ from each other and from other kinds of seeds? What part do seeds play in a plant's life cycle?

Science in a Snap! Animal Observations

Think about animals you have **observed.** Perhaps you have a pet, or maybe you have watched birds, squirrels, or other animals outdoors. Make a list of animal behaviors you have observed. Tell which behaviors you think the animals were born knowing how to do. Which behaviors were learned?

Science in a Snap! Identify Plants That People Eat

In 30 seconds, list as many foods as you can think of that come from plants. **Compare** lists with a partner. Cross off any foods that are on both lists. Which foods are on your list but not on your partner's list? People eat plants to obtain energy. Where do most plants get their energy?

Science in a Snap! Observe Different Kinds of Leaves

Observe the 2 plants. One plant has a thick stem and thick leaves. The other plant has a thin stem and thin leaves. Obtain a leaf from each plant. Use scissors to cut each leaf in half. Which plant do you think will survive better in a dry environment? Why might some plants survive better in dry conditions than others?

Science in a Snap! Analyze Environmental Impact

Observe the picture. Identify the living and nonliving parts shown in this environment. **Analyze** how the living and nonliving parts of the environment interact. How do plants and animals have an impact on this environment? How do humans have an impact on this environment?

Science in a Snap! Observe How Eyes React to Light

Your teacher will dim the lights in the room. Wait 2 minutes until your eyes adjust to the new light level in the room. Then **observe** your partner's eyes. Look carefully at the pupils. Continue to observe your partner's eyes as your teacher turns on the lights again. How did your partner's pupils change after the lights became brighter? Why do you think this happened?

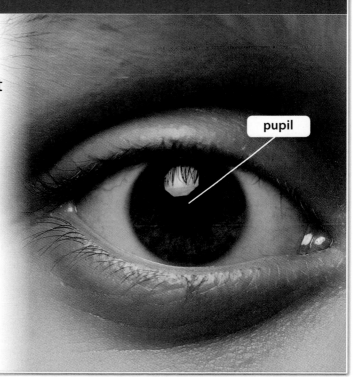

pupil

Investigate a Plant Life Cycle

**Question** What are the stages in the life cycle of a radish plant?

Science Process Vocabulary

observe verb

When you **observe,** you use your senses to learn about an object or event.

compare verb

When you **compare,** you tell how objects or events are alike and different.

How are these plants alike and different?

Materials

spoon

3 radish seeds

cup with soil

spray bottle with water

hand lens

brush

Purple Coneflower Life Cycle diagram

What to Do

1 Use the spoon to plant the 3 radish seeds in the soil. Make sure that soil covers each seed. Moisten the soil using the water in the spray bottle. Place the cup in a sunny place.

2 **Observe** the cup each day. Use the spray bottle to keep the soil moist but not wet. Record your observations in your science notebook.

3 When the seeds sprout, observe and draw the plant parts. Continue to observe and draw the plants over time, noting how they change. Record your observations.

What to Do, continued

4 When flowers appear on the plants, use a hand lens to examine the parts of each flower. Record your observations and draw the flowers.

5 Use the brush to move the pollen between the flowers. Swirl the brush around in an open flower so that the brush picks up pollen. Then swirl the brush inside another flower so that the pollen sticks to the inside of the flower. Repeat for all open flowers. Moving pollen from flower to flower will allow seeds to form.

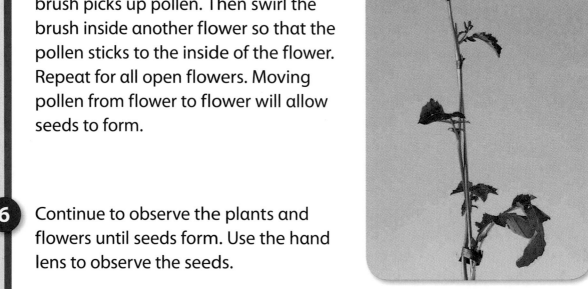

6 Continue to observe the plants and flowers until seeds form. Use the hand lens to observe the seeds.

7 Use your observations to draw the life cycle of the radish plant in your science notebook. Use the Purple Coneflower Life Cycle diagram to **compare** the life cycle of the radish with the life cycle of the purple coneflower.

Record

Write and draw in your science notebook.
Use a table like this one.

SCIENCE notebook

Radish Plants

Date	Observations and Drawings

Explain and Conclude

1. What changes did you **observe** in the plants as they grew?

2. How are the life cycles of the radish and purple coneflower alike and different?

Life Cycle of the Radish Plant

These purple coneflower plants will grow and change as they complete their life cycles.

Math in Science

Recording Observations

Making a diagram during an investigation is an important way that scientists keep accurate records of their observations. Scientists use different kinds of diagrams.

Table and Graph Diagrams Sometimes scientists record observations as tables with numbers. This kind of diagram is especially helpful when scientists make measurements.

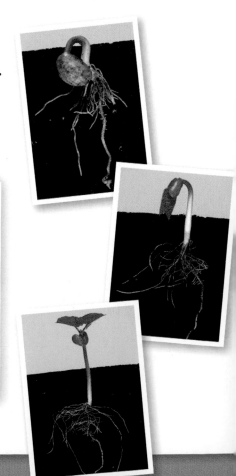

Plant Growth

Day	Plant Height (cm)
1	0
3	1
5	3
7	5
9	8

Sometimes scientists graph the data from their tables.

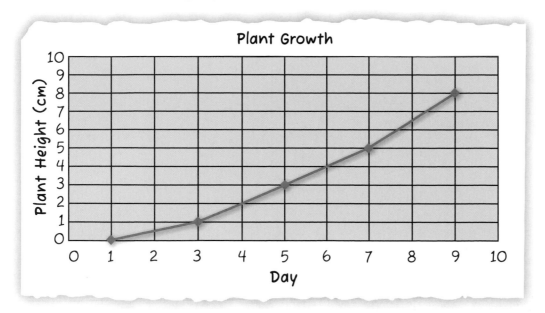

Scale Diagrams Often when scientists make a drawing of an object that is very large or very small, they compare each part of the object to the size of the real thing. Sometimes they make actual measurements. Then they draw the parts so that the relative sizes remain the same.

A scientist observed a leaf through a microscope. She saw tiny openings in the leaf. When the scientist drew what she saw, she made the parts larger but she drew them to scale.

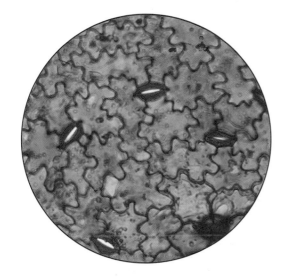

View of leaf through a microscope

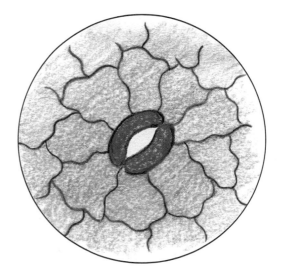

Scientist's drawing of what she saw

Sequence Diagrams Scientists might want to show the order in which something happens, such as a life cycle. A sequence diagram is useful for this purpose. The sequence diagram shows how the bald cypress tree grows from a seed to an adult tree.

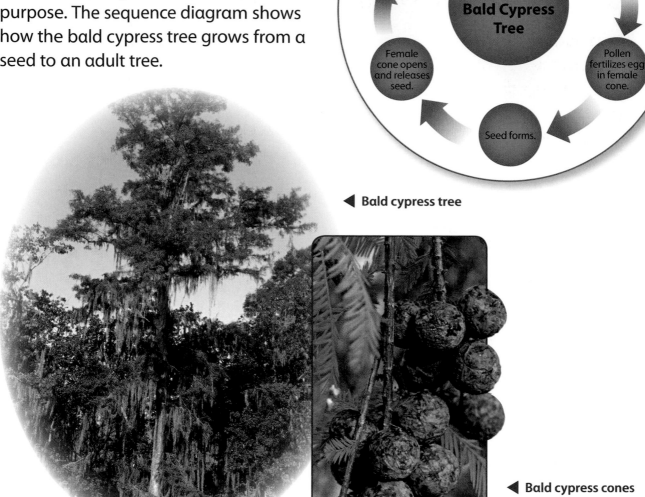

Life Cycle of Bald Cypress Tree

- Tree matures.
- Pollen moves from male cone to female cone.
- Pollen fertilizes egg in female cone.
- Seed forms.
- Female cone opens and releases seed.
- Seed germinates.

◀ Bald cypress tree

◀ Bald cypress cones

SUMMARIZE

What Did You Find Out?

1 Why do scientists use diagrams?

2 What are three kinds of diagrams? What kind of information does each diagram show?

Practice Reading a Diagram

The diagram shows the life cycle of an orange tree. Study the diagram. Then compare the life cycle of the bald cypress tree and the life cycle of the orange tree.

Life Cycle of Orange Tree

- Tree matures.
- Pollen moves from anther to stigma.
- Pollen fertilizes egg in ovary.
- Seed forms.
- A fruit develops around the seed.
- Seed germinates.

◀ Orange tree flower

Investigate Flowers

Question What are the parts of a flower?

Science Process Vocabulary

compare verb

To **compare** objects or events, tell how their characteristics are alike and different.

These flowers both have petals, but the petals have different shapes.

infer verb

When you **infer,** you use what you know and what you observe to draw a conclusion.

I infer that the color of the flower attracts insects.

Materials

flower

Parts of a Flower Diagram

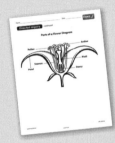
scissors

hand lens

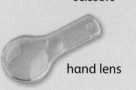
ruler

microscope and slide

white paper

What to Do

1 **Observe** the flower. Identify the following parts: petal, anther, stamen, ovary, and pistil. Use the Parts of a Flower Diagram to help you.
Draw the whole flower and label each part in your science notebook. Make sure you draw the parts to scale.

2 Use the scissors to carefully cut a stamen from the flower. Observe the stamen with the hand lens. **Measure** its length with the ruler. Record your observations.

3 Remove another stamen from the flower and gently tap it on the microscope slide. The powdery material on the slide is pollen. Observe the pollen under the microscope. Record your observations and draw the pollen.

What to Do, continued

4 Place the stamen on the sheet of white paper and label it.

5 Remove the other flower parts with the scissors. Measure each part and observe it with the hand lens or under the microscope. Record your observations. Place each part of the flower on the paper and label it.

6 **Compare** your plant parts with those of other groups. Identify ways that your flowers are similar and ways that they are different.

Record

Write and draw in your science notebook.
Use a table like this one.

Parts of a Flower

Part	Observations	Length (cm)
Stamen		
Petal		

Pollen Under the Microscope

Whole Flower

Explain and Conclude

1. What flower parts did you **observe?**

2. Describe the pollen. **Infer** how its structure might be useful during the processes of pollination.

3. What similarities and differences did you find when you **compared** your flower parts with those of other groups?

Think of Another Question

What else would you like to find out about the parts of a flower? How could you find an answer to this new question?

These plants reproduce with flowers.

Investigate Earthworm Behavior

Question Will an earthworm move toward a light or dark environment?

Science Process Vocabulary

predict verb

When you **predict,** you tell what you think will happen.

I predict the earthworm will dig into the soil.

conclude verb

You **conclude** when you use information, or data, from an investigation to come up with a decision or answer.

I will use my observations to conclude which environment an earthworm will move toward.

Materials

plastic pan

2 paper towels

spray bottle with water

2 sheets of black construction paper

tape

spoon

3 earthworms

flashlight

stopwatch

28

What to Do

1 Lay the paper towels on the bottom of the plastic pan. Use the spray bottle to moisten the paper towels. Do not make them so wet that puddles form.

2 Cover half of the plastic pan with black construction paper. Use tape to hold the construction paper in place.

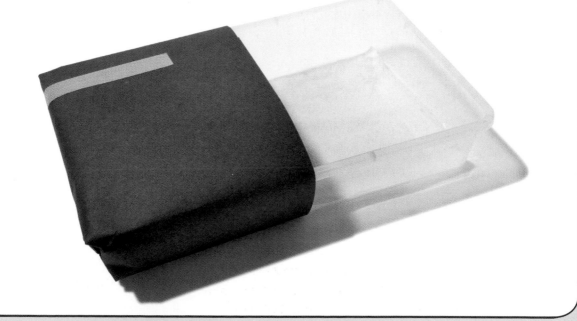

What to Do, continued

3 Use the spoon to place 3 earthworms in the center of the pan. To which part of the pan will the earthworms move? Write your **prediction** in your science notebook.

4 Turn off the classroom lights. Shine the flashlight directly over the uncovered side of the pan. Use the stopwatch to time 5 minutes. Then **observe** the location of the earthworms. Record your observations.

5 Repeat steps 3 and 4 three more times. Record your observations.

Record

Write in your science notebook.
Use a table like this one.

Earthworm Behavior

Trial	Prediction	Observations	
		Number of Worms in the Dark After 5 Minutes	Number of Worms in the Light After 5 Minutes
1			
2			

Explain and Conclude

1. Do your results support your **predictions?** Explain.

2. **Compare** the behavior of all 3 earthworms. What do you think caused any differences?

3. Use the results of your **investigation** to draw a **conclusion** about earthworm behavior. Do you think the earthworms inherited this behavior or was this behavior affected by the environment? Explain.

Think of Another Question

What else would you like to find out about whether earthworms will move toward light or dark environments? How could you find an answer to this new question?

Earthworms are affected by changes in their environment.

Investigate Insect Life Cycles

Question How does the life cycle of a butterfly compare with the life cycle of a grasshopper or dragonfly?

Science Process Vocabulary

observe verb

Scientists use tools to **observe** things that are too small to see with their eyes alone.

compare verb

When you **compare,** you tell how objects and events are alike and different.

Materials

caterpillars

hand lens

butterfly habitat

Insect Life Cycle diagrams

The colors of the butterflies' wings are the same, but they can be different sizes.

What to Do

1 **Observe** the photo of the butterfly egg. Draw the egg and record your observations in your science notebook.

Butterflies reproduce by laying eggs. The eggs in this photograph appear larger than actual eggs, which are about the size of the head of a pin.

2 A caterpillar is the larva stage of a butterfly life cycle. Use the hand lens to observe the larvae in the container. Record your observations and draw the larvae.

3 Observe the larvae each day. Record any changes you observe. When pupae form, observe a pupa with a hand lens. Record your observations and draw the pupa.

What to Do, continued

4 Your teacher will pin the paper circle with the pupae to the butterfly habitat. Observe the pupae each day. Record your observations.

5 When the pupae become adults, observe and draw the butterflies. Record your observations.

6 Use your observations and drawings to make a diagram of the life cycle of the butterfly. Label the stages **Egg, Larva, Pupa,** and **Adult.**

7 Choose one of the Insect Life Cycle diagrams. **Compare** the life cycle in the diagram you chose with the butterfly life cycle you drew.

8 **Share** your observations of the diagrams with other groups. Did you observe the same similarities and differences between the insect life cycles?

Record

Write and draw in your science notebook.
Use a table and diagram like these.

Butterfly Growth

Date	Stage	Observations and Drawings

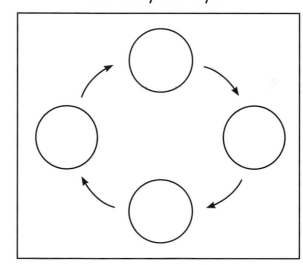

Butterfly Life Cycle

Explain and Conclude

1. What major changes did you **observe** as the butterfly went through its life cycle?

2. How did the life cycle of the butterfly **compare** to the life cycle of the grasshopper or dragonfly? Which insect undergoes complete metamorphosis? Which undergoes incomplete metamorphosis?

Think of Another Question

What else would you like to find out about the life cycles of butterflies, grasshoppers, and dragonflies?

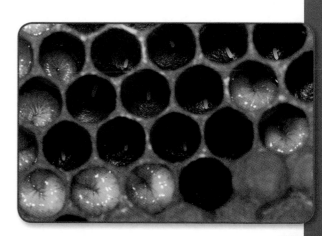

These honeybee larvae will change into pupae before becoming adult honeybees.

Investigate Owl Pellets

Question What objects can you find in an owl pellet?

Science Process Vocabulary

count verb

When you **count,** you tell the number of something.

> I count 4 bones that are alike.

data noun

Data are observations and information that you collect and record in an investigation.

> I will record these data in my science notebook.

Materials

safety goggles   protective gloves

owl pellet   2 paper plates

hand lens   craft stick

forceps   Bone Sorting Chart

6 resealable bags   tape

What to Do

1 Owls usually swallow their food whole. They digest some parts of the food. They eject the other parts through their mouths in the form of pellets. Pellets may contain bones, fur, teeth, and other animal parts.

2 Put on your safety goggles and gloves. Unwrap the owl pellet and place it on a paper plate. Use the hand lens to **observe** the owl pellet. Record your observations in your science notebook. **Predict** how many different animal parts you will find in the pellet.

3 Hold the owl pellet in place with the craft stick while you gently pull apart the pellet with the forceps. Use the forceps to remove pieces of bone, teeth, and other materials.

What to Do, continued

4 Place the bones, teeth, and other materials on a second paper plate. Use a hand lens to observe the materials. Roll any pieces of fur between your fingers to find small bones or teeth. Separate all the bones from other materials.

5 Look closely at the shape of each bone. Use the Bone Sorting Chart to help you sort the bones by bone type.

6 **Count** and record the number of bones of each type. Then graph the **data.**

7 Carefully place the bones of each type in a resealable bag. Label the bags with the bone type. You will use the bones again in the next investigation.

Record

Write in your science notebook.
Use a table and graph like these.

Owl Pellet

Object	Observations	How Many?
Whole pellet		1
Skulls		

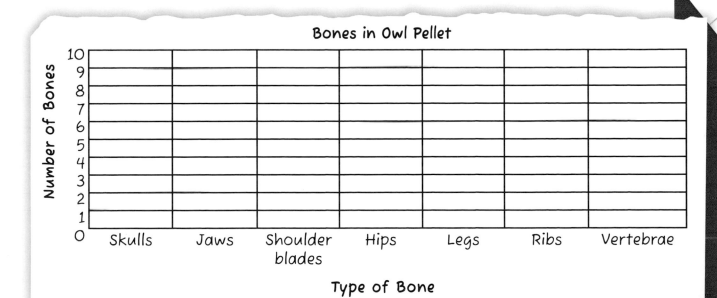

Bones in Owl Pellet

Number of Bones — Type of Bone: Skulls, Jaws, Shoulder blades, Hips, Legs, Ribs, Vertebrae

Explain and Conclude

1. What materials did you remove from the owl pellet? How do these materials help you trace the energy the owl gets from its food back to the sun?

2. Based on the **data** in your graph, **estimate** how many different animals are represented by the bones in the pellet. Explain.

3. How might you use owl pellets to make **inferences** about what owls eat?

Think of Another Question

What else would you like to find out about the objects found in owl pellets? How could you find an answer to this new question?

Investigate Food Chains and Webs

Question How can you use an owl pellet to infer a food chain and food web?

Science Process Vocabulary

share verb

When you **share** results, you tell or show what you have learned.

conclude verb

You **conclude** when you use data from an investigation to come up with a decision or an answer.

> I conclude that these two bones came from the same kind of animal.

Materials

protective gloves

bags with bones

forceps

index cards

Bone Sorting Chart

Skeleton diagrams

paper

construction paper

glue

Food Web diagram

What to Do

1 Put on the plastic gloves. Use the forceps to carefully remove the bones from each resealable bag and place them on an index card. Make sure to keep the bones from each bag together.

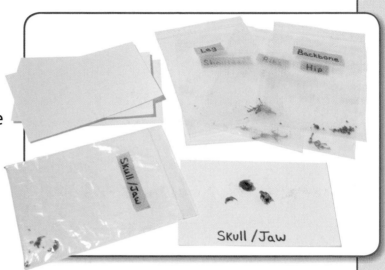

2 Use the Bone Sorting Chart to identify the animal to which each bone belongs. If you have bones from more than one animal, sort all the bones from each animal and place them on separate sheets of paper. Choose the animal for which you have the most bones. Place the bones from the other animals into a resealable bag and set the bag aside.

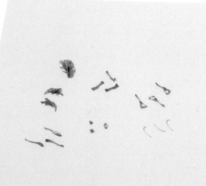

What to Do, continued

3 Use the Skeleton diagram for the type of animal you found to make a skeleton on construction paper. Glue the bones to the paper. Draw in any missing bones. Label the skeleton with the name of the animal.

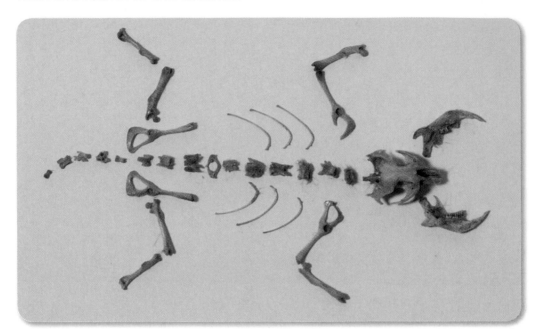

4 Use the results of the investigation and the Food Web diagram to draw a food chain that includes the owl and the animal whose skeleton you put together.

5 **Share** your skeleton and food chain with other groups. **Compare** the food chains that include different animals. Construct a class food web that includes all the animals identified in the owl pellets.

Record

Write and draw in your science notebook.

My Food Chain

Class Food Web

Explain and Conclude

1. Is the animal whose skeleton you assembled a producer or a consumer? Where does the owl fit in the food chain?

2. How many different kinds of skeletons did your class find in the owl pellets? Which kind of animal skeleton was found the most?

3. What can you **conclude** about the types of animals that owls eat?

Think of Another Question

What else would you like to find out about using owl pellets to infer about food chains and food webs? How could you find an answer to this new question?

Great Horned Owl
(Bubo virginianus)

Investigate Plant Fossils

Question How can you use plant fossils to infer what some environments were like long ago?

Science Process Vocabulary

compare verb

When you **compare,** you tell how objects or events are alike and different.

How are the ferns alike and different?

infer verb

When you **infer,** you use what you know and what you observe to draw a conclusion.

I can use my observations of the fossil to infer what its environment was like.

Materials

plant fossil

hand lens

tissue paper

pencil

Plant Information chart

additional materials

44

What to Do

1 Choose a fossil to study. **Observe** the fossil with the hand lens. Examine its shape. Which plant parts can you observe in your fossil? Record your observations in your science notebook.

2 Place a piece of tissue paper over your fossil. Gently rub your pencil back and forth across the tissue paper. You will see the shape of your fossil on the tissue paper.

Rub with the side of your pencil tip to avoid ripping the tissue paper.

What to Do, continued

3 **Compare** your fossil rubbing to the photos of the leaves of the sword fern, sycamore, and elm. Which modern plant looks the most like your fossil? Record your observations.

sword fern

4 Read about the modern plant that looks like your fossil on the Plant Information chart. Use the information on the chart to **infer** what the environment was like where your fossil plant lived.

sycamore

5 With your group, design a display that shows the environment of your fossilized plant. **Share** your display with the class and discuss your inferences.

elm

Record

Write and draw in your science notebook.
Use a table like this one.

SCIENCE notebook

Fossil Observations

Shape	
Plant parts	
Other observations	
Which modern plant looks the most like your fossil?	

Explain and Conclude

1. How was your fossil plant similar to the modern plant on the chart? How was it different?

2. **Compare** your fossil with the fossils of other groups. How are they alike and different?

3. **Infer** what the environment was like at the time your fossil plant lived. Explain your inference.

Think of Another Question

What else would you like to find out about using plant fossils to learn about environments from long ago? How could you find an answer to this new question?

Many different types of ferns live in this forest.

Investigate Animal Survival

Question How can an animal's color help it survive in its environment?

Science Process Vocabulary

compare verb

When you **compare,** you tell how objects are alike and different.

infer verb

When you **infer,** you use what you know and what you observe to draw a conclusion.

I infer that the insect's color may prevent birds from seeing it very well.

Materials

6 pieces of construction paper

hole punch   paper cup

white cloth   stopwatch

patterned cloth   green cloth

What to Do

1 Use a hole punch to punch out 10 paper dots from each piece of construction paper. Each dot stands for one animal. Place the dots in the paper cup.

2 Place the white cloth on the table. The cloth is a **model** habitat where the animals live. Without letting your partner see, spread out the paper dots on the white cloth.

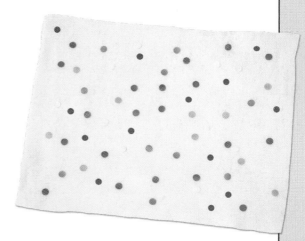

3 Your partner will represent a predator. When you say "Go," your partner should begin removing as many of the dots from the cloth as possible. He or she should pick up the dots one at a time. Use the stopwatch to time your partner for 15 seconds. Then tell him or her to "Stop."

What to Do, continued

4 **Count** the total number of dots of each color that your partner collected. Record the **data** in your science notebook.

5 Place all the dots on the white cloth again. Do steps 3 and 4 two more times.

6 Replace the white cloth with either the patterned cloth or the green cloth. Repeat steps 2–4 three times with the new cloth.

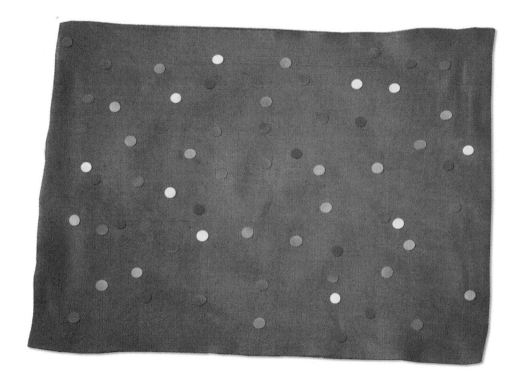

Record

Write in your science notebook. Use tables like these.
Write the colors of the dots and the cloth you used.

SCIENCE notebook

Dots Removed From White Cloth

Trial	_____ Dots	_____ Dots	_____ Dots	_____ Dots	_____ Dots	_____ Dots
1						

Dots Removed From _____ Cloth

Trial	_____ Dots	_____ Dots	_____ Dots	_____ Dots	_____ Dots	_____ Dots
1						

Explain and Conclude

1. **Compare** the number of dots of each color picked up on the white cloth and on the other cloth. Explain any differences.

2. Use the results of this **investigation** to **infer** how blending in with its environment might help an animal survive.

Think of Another Question

What else would you like to find out about how an animal's color helps it survive? How could you find an answer to this new question?

This octopus is difficult to see because of its color.

Investigate Plants and Water

Question How can a plant affect the water in its environment?

Science Process Vocabulary

observe verb

When you **observe,** you use your senses to learn about an object or event.

I will observe the plant over time to see what happens if it is not watered.

infer verb

When you **infer,** you use what you know and what you observe to draw a conclusion.

I infer that these plants died from lack of water.

Materials

safety goggles

plant cutting

hand lens

index card with waxed paper

pencil

clay

plastic cup with water

plastic cup

black marker

What to Do

1 Put on your safety goggles. Use the hand lens to **observe** the leaves of the plant cutting. Record your observations in your science notebook.

2 Use the pencil to make a small hole in the center of the index card and waxed paper. Hold the index card so that the waxed paper is on the bottom. Very carefully pull the stem of the plant cutting through the hole. Seal around the opening with clay.

What to Do, continued

3 Place the plant cutting in the cup of water so the index card rests on top. Cover the top of the plant with the other plastic cup. Observe the plant and the water in the bottom cup. Record your observations. Mark the water level by drawing a line on the bottom cup.

Make sure that the stem of the plant is in the water.

4 Place the plant in a sunny spot.

5 Wait 24 hours and observe the plant and cups. Be sure to observe the water level in the bottom cup. Record your observations.

6 Observe the plant again after 48 hours. Record your observations and **compare** them to your observations from the previous day.

Record

Write and draw in your science notebook.
Use a table like this one.

Plant Cutting and Cups

Time	Observations
Plant only	
Plant in water: start	
Plant in water: 24 hours	

Explain and Conclude

1. What did you **observe** in the cups after 24 hours? What did you observe after 48 hours?

2. **Infer** what happens to some of the water that plants take in.

3. Use the results of this **investigation** to infer how plants might affect groundwater. How might they affect water in the air?

Think of Another Question

What else would you like to find out about how plants affect water in the environment? How could you find an answer to this new question?

Trail through forest, Potsdam, Germany

In a year, a tall tree can release over 100,000 liters of water into the air.

Investigate Colors in Green Leaves

Question What colors are present in a green leaf?

Science Process Vocabulary

investigate verb

You **investigate** when you make a plan to answer a question and carry it out.

I will investigate to find out what colors are present in green leaves.

infer verb

When you **infer,** you use what you know and what you observe to draw a conclusion.

I can use my results to infer about the colors in a green leaf.

Materials

green leaves

safety goggles

hand lens

scissors

2 plastic cups

spoon

rubbing alcohol

graduated cylinder

stopwatch

coffee filter

metric ruler

2 craft sticks

tape

paper towel

Do an Experiment

Write your plan in your science notebook.

Make a Hypothesis

In this investigation, you will separate the colors in a green leaf. What colors will you observe? Write your **hypothesis.**

Identify, Manipulate, and Control Variables

Which variable will you change?
Which variable will you observe or measure?
Which variables will you keep the same?

What to Do

1   Put on your safety goggles. Choose a leaf. **Observe** the leaf with a hand lens. Record your observations in your science notebook.

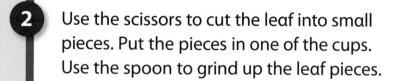

2   Use the scissors to cut the leaf into small pieces. Put the pieces in one of the cups. Use the spoon to grind up the leaf pieces.

What to Do, continued

3 Use the graduated cylinder to add 25 mL of alcohol to the cup. Let the leaves and the rubbing alcohol sit for 20 minutes. Observe and record the color of the solution in the cup.

4 Pour 25 mL of rubbing alcohol into the second cup. Do not add anything else to this cup. Observe and record the color of the alcohol.

5 Cut 2 strips from a coffee filter. Each should be about 3 cm wide and 9 cm long. Cut a point at one end of each strip. Tape each strip around a craft stick. Place each craft stick across the top of one of the cups. The points of the strips should just touch the alcohol in each cup. Observe the coffee filter strip in each cup and record your observations.

6 After 1 hour, remove the strips from the cups and place them on a paper towel. Observe the strips after they have dried. Record your observations.

Record

Write or draw in your science notebook.
Use a table like this one.

Observations of Cups

	Color of Liquid	Coffee Filter Strip: Start	Coffee Filter Strip: Dried
Leaf solution			
Alcohol			

Explain and Conclude

1. What color was the solution in the cup with the leaf pieces? **Compare** your **observation** of the leaf solution and the rubbing alcohol with no leaves.

2. What colors did you observe on each coffee filter strip after 1 hour? **Infer** where the colors on the strip came from. Explain your reasoning.

3. Shorter daylight hours and cooler temperatures cause some trees to stop making the material that gives leaves their green color. **Infer** how the leaves change color.

Think of Another Question

What else would you like to find out about colors in leaves?
How could you find an answer to this new question?

What colors are present in the leaves of these trees?

Investigate How Your Brain Can Work

Question What happens when you identify different colors of letters in words?

Science Process Vocabulary

data noun

Data are observations and information you collect during an investigation. One kind of data is the amount of time it takes for something to happen.

> I collected data about the amount of time it took to read a list of words.

compare verb

When you **compare,** you tell how objects or events are alike and different.

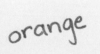

Materials

2 sheets of white paper

markers

stopwatch

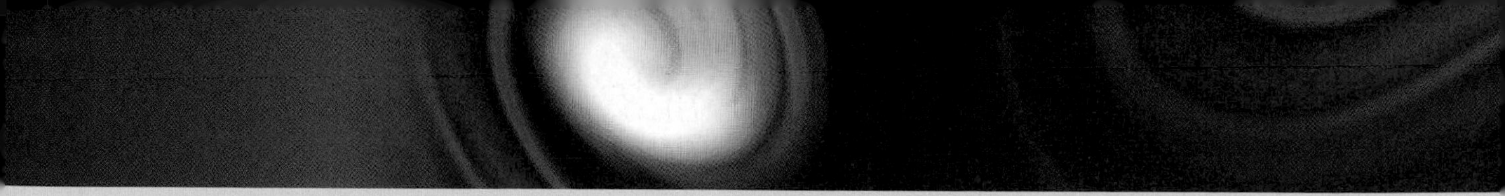

What to Do

1 On a sheet of paper, write the following words in 3 rows of 4 words, as shown in the photo: **red, yellow, orange, green, blue, purple, black, brown.** You can put the words in any order and repeat words to get a total of 12 words. Use the markers to write each word in the color that matches it. For example, the word **red** should be written using the red marker. Label the paper **Test A.**

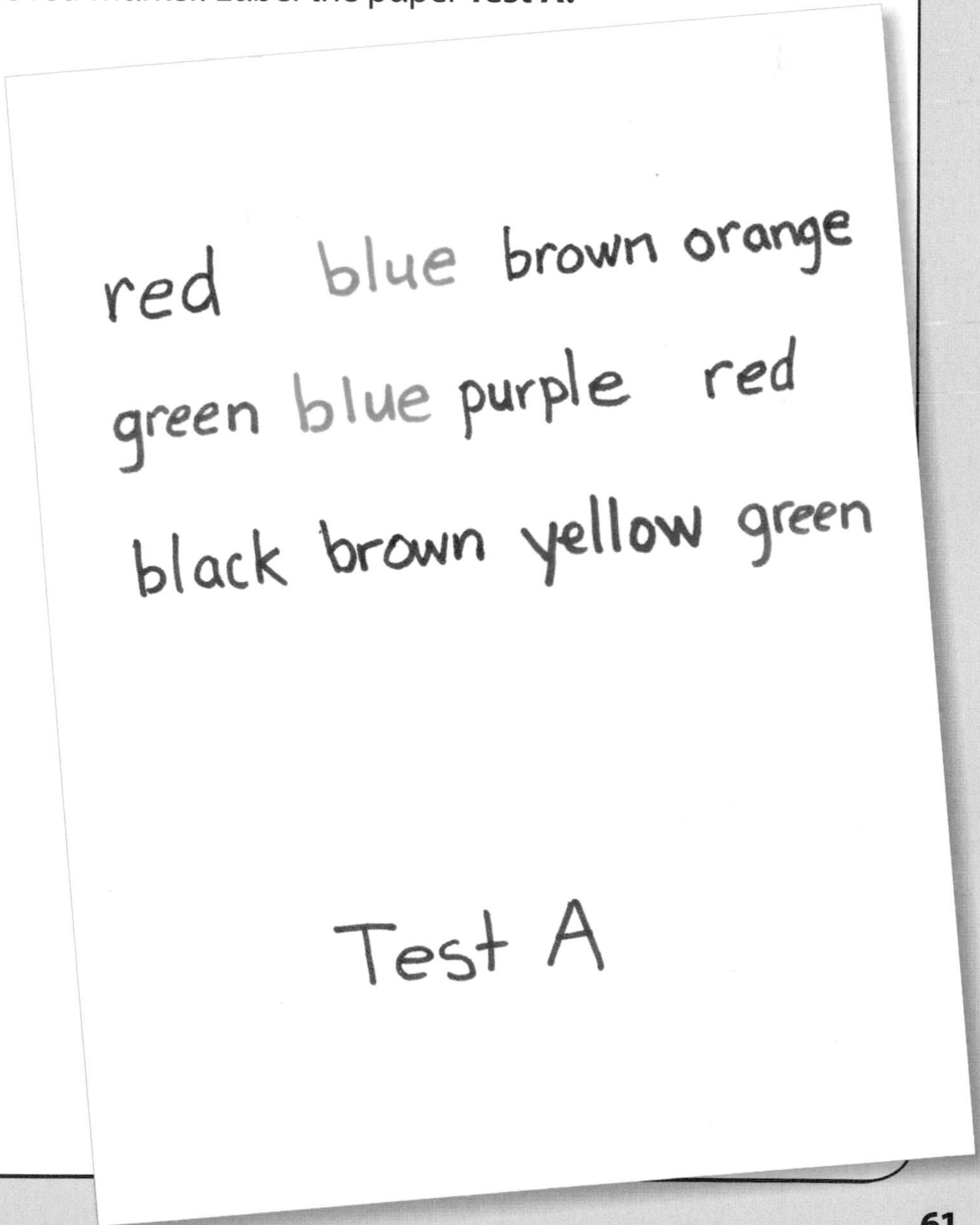

red   blue brown orange

green blue purple red

black brown yellow green

Test A

What to Do, continued

2 On the second sheet of paper, write the words again, in the same order you wrote them the first time. This time use a different color for each word. For example, you may choose to write the word **brown** using the blue marker. Label the paper **Test B.**

red blue brown orange
green blue purple red
black brown yellow green

Test B

3 Use the stopwatch to time your partner as he or she correctly names the COLOR of the letters of each word on Test A. Record the **data.**

4 Then time your partner as he or she correctly names the COLOR of the letters of each word on Test B. For example, if the word black is written in purple marker, say "Purple." Record the data.

5 Switch roles with your partner and repeat steps 3 and 4.

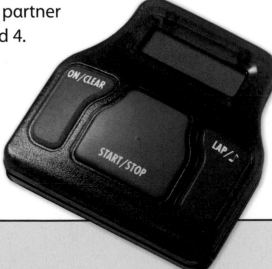

Record

Write in your science notebook.
Use a table like this one.

	Identifying Colors	
Who?	Time to Read Test A (s)	Time to Read Test B (s)
Me		
Partner		

Explain and Conclude

1. **Compare** the times needed to complete Test A and Test B. Which test did you take longer to finish? Why do you think that is so?

2. Compare your times with your partner's times. Explain any differences.

3. What body parts did you use as you took these tests? How did each part help you with the task?

Think of Another Question

What else would you like to find out about how the brain processes information? How could you find an answer to this new question?

Tests like this can help determine how a person sees different colors.

Investigate Exercise and Heart Rate

Question How does exercise affect heart rate?

Science Process Vocabulary

count verb

When you **count,** you tell the number of something.

> I can count the number of times my heart beats in a minute by checking my pulse.

stopwatch

conclude verb

When you **conclude,** you decide something based on evidence.

> I conclude that a change in activity can change how fast my heart beats.

Do an Experiment

Write your plan in your science notebook.

Make a Hypothesis

In this investigation, you will choose an exercise and measure your partner's heart rate before and after he or she does the exercise. How will exercise affect heart rate? Write your **hypothesis.**

Identify, Manipulate, and Control Variables

Which variable will you change?
Which variable will you observe or measure?
Which variables will you keep the same?

What to Do

1. Work with a partner. Choose an exercise you will do in this **investigation.** You may choose running in place, jumping jacks, or sit-ups.

2. Have your partner sit on a chair. Find your partner's pulse by placing your first 2 fingers on the thumb-side of his or her wrist, below the bottom of the thumb. Press firmly with flat fingers. Slowly move your fingers around until you feel the beats. Practice **counting** the beats until you feel comfortable doing it.

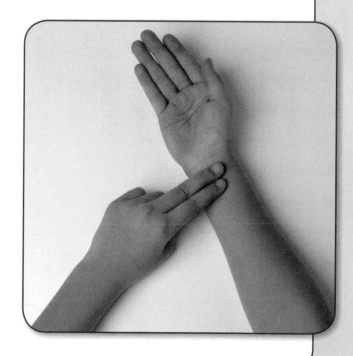

What to Do, continued

3 **Measure** your partner's resting pulse by counting the beats for 1 minute. Use the stopwatch to time yourself. Record the **data** in your science notebook.

4 Have your partner do the exercise you chose for 1 minute. As soon as your partner stops, measure his or her pulse for 1 minute. Record the data.

5 Wait 3 minutes. Then check your partner's pulse to make sure it has returned to the resting rate. Repeat step 4 two more times.

6 Switch roles with your partner and repeat steps 2–5.

Record

Write in your science notebook.
Use a table like this one.

Effects of Exercise on Heart Rate

Who?	Kind of Exercise	Resting Heart Rate (beats/min)	Heart Rate After Exercising (beats/min)
Me, trial 1			
Me, trial 2			

Explain and Conclude

1. **Compare** your heart rate before and after exercising. What can you **conclude** about how exercise affects heart rate?

2. Compare your results with groups that did the same exercise. Did they get similar results? Explain any differences.

3. Compare your results with groups that did different exercises. Did one kind of exercise affect heart rate more than other kinds? Why do you think that is so?

Think of Another Question

What else would you like to find out about how exercise affects heart rate? How could you find an answer to this new question?

How does running affect heart rate?

Do Your Own Investigation

 **Choose one of these questions, or make up one of your own to do your investigation.**

- How is a plant's growth affected by the amount of water it receives?
- How does temperature affect how fast a mealworm grows?
- If I place pill bugs in a habitat with cereal, cornmeal, and green leaves, which food will they eat most?
- How does the shape of a model fish affect how it moves through water?
- How do earthworms affect the growth of lima bean plants?
- How does practice affect the amount of time it takes to stack 10 centimeter cubes?

Science Process Vocabulary

measure verb

When you **measure,** you find out how much or how many.

I can measure the plant's height with a ruler.

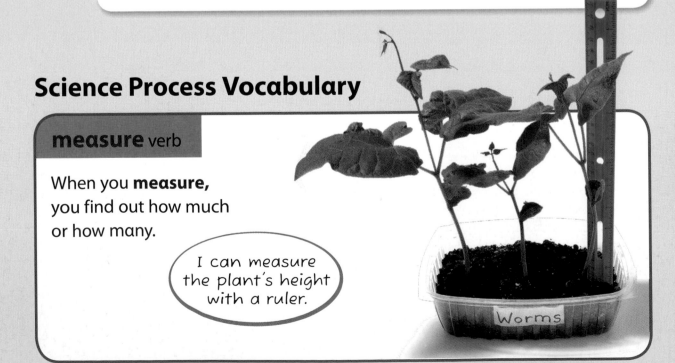

Open Inquiry Checklist

Here is a checklist you can use when you investigate.

☐ Choose a **question** or make up one of your own.

☐ Gather the materials you will use.

☐ If needed, make a **hypothesis** or a **prediction.**

☐ If needed, identify, manipulate, and control **variables.**

☐ Make a **plan** for your **investigation.**

☐ Carry out your **plan.**

☐ Collect and record **data. Analyze** your data.

☐ Explain and **share** your results.

☐ Tell what you **conclude.**

☐ Think of another question.

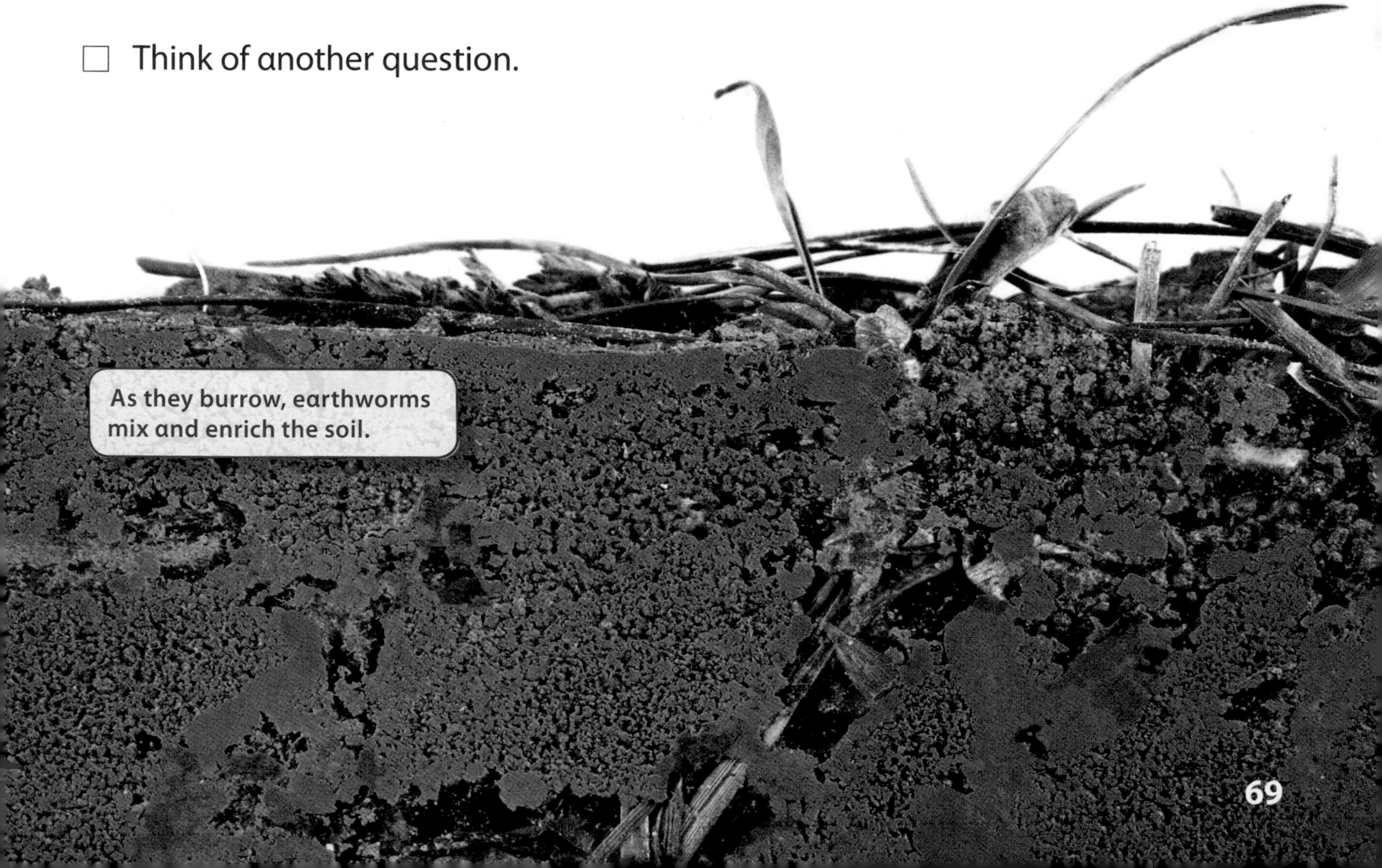

As they burrow, earthworms mix and enrich the soil.

Write Like a Scientist

Write About an Investigation

How Animals Get Energy

The following pages show how one student, Taj, wrote about an investigation. As he observed some pill bugs in his yard, Taj became interested in how the pill bugs get energy. He decided to do an investigation to determine what kinds of food pill bugs eat. Here are some things Taj thought about before he started his investigation:

Pill bugs are often found under decaying leaves.

- Taj knew that pill bugs were easy and safe to keep, so he decided that he could easily use them for his investigation.

- He would build a habitat for the pill bugs and then place several pill bugs in the habitat.

- He would give the pill bugs a variety of foods.

- He would observe which foods the pill bugs eat to see if they eat more of one kind of food.

Model

Question

If I place pill bugs in a habitat with cereal, cornmeal, and green leaves, which food will they eat most?

> State the question so it is clear what you are trying to find out.

Materials

plastic container

soil

spray bottle with water

cereal flakes

cornmeal

green leaf

10 pill bugs

plastic wrap

tape

> Make sure you have all of the materials before you begin your investigation.

Your Investigation

Now it's your turn to do your investigation and write about it. Write about the following checklist items in your science notebook.

☐ **Choose a question or make up one of your own.**

☐ **Gather the materials you will use.**

Model

My Hypothesis

If I give pill bugs cornmeal, cereal flakes, and leaves, then the pill bugs will eat mostly leaves because leaves are part of their natural environment. The pill bugs might also eat a little of the cornmeal and cereal.

Taj based his hypothesis on the fact that in their natural habitat, pill bugs eat decaying leaves and other vegetation.

Your Investigation

☐ **If needed, make a hypothesis or prediction.**

Write your hypothesis or prediction in your science notebook.

Model

Variable I Will Change

The pill bugs will have 3 different kinds of food: cornmeal, cereal flakes, and leaves.

Variable I Will Observe or Measure

I will observe the pill bugs' behavior to see whether more pill bugs go to a particular kind of food. I will also observe how much food is left.

Variables I Will Keep the Same

Everything else will be the same. I will use the same foods and the same amounts of food for each trial. I will use the same number of pill bugs. They will be placed the same distance from the foods. I will observe for the same amount of time for each trial.

> Answer these 3 questions:
> 1. What one thing will I change?
> 2. What will I observe or measure?
> 3. What things will I keep the same?

 **Your Investigation**

☐ **If needed, identify, manipulate, and control variables.**

Write about the variables in your investigation.

Model

My Plan

1. Cover the bottom of a plastic container with soil. Use the spray bottle to make the soil moist.

2. Put 2 spoonfuls each of cereal flakes and cornmeal in the habitat. The piles should be in the corners of the container.

3. Place a pile of leaf pieces in the habitat across from the cereal and cornmeal. The 3 piles should form a triangle. All piles should be about the same size.

4. Place 10 pill bugs in the middle of the habitat.

5. Cover the habitat with plastic wrap. Tape the plastic wrap to the container. Put some holes in the plastic wrap.

6. Observe the pill bugs every 15 minutes for 1 hour. Count the number of pill bugs that are at each pile of food. Also observe how much food is left in each pile after 1 hour.

7. Repeat step 6 every day for 3 days.

8. Do 2 more trials. Use different pill bugs for each trial.

SCIENCE my notebook

Your Investigation

☐ **Make a plan for your investigation.**

Write the steps for your plan.

Your plans should be clear and detailed. Another student should be able to follow your steps.

Model

I carried out the 8 steps of my plan.

Make notes about any changes you made to the plan. Taj did not have any changes to make.

Your Investigation

☐ **Carry out your plan.**

Be sure to follow your plan carefully.

Model

Number of Pill Bugs at Different Foods

Time		Cornmeal	Cereal Flakes	Leaves	Observations
Day 1	15 min.	none	none	5	
	30 min.	2	none	3	
	45 min.	1	3	6	
	1 hour	1	2	4	Most of the cornmeal and cereal is left. Not as many leaves are left.
Day 3	15 min.	2	2	4	
	30 min.	none	none	7	
	45 min.	none	3	5	
	1 hour	1	1	6	Not many leaves are left. Some cornmeal and cereal is left.

My Analysis

The pill bugs ate a little of the cornmeal and cereal flakes. They ate most of the leaves.

> Examine your data carefully to understand what happened.

 **Your Investigation**

☐ **Collect and record data. Analyze your data.**

Collect and record your data, and then write your analysis.

How I Shared My Results

I presented my results to the class. First, I described how I set up the investigation. Then, I shared my observations and my analysis. Finally, I explained how pill bugs get energy from their environment.

A presentation is one way to share your results with others.

My Conclusion

The pill bugs ate mostly leaves and mostly stayed at the pile of leaf pieces. This makes sense, because leaves are available in a pill bug's natural habitat. The results support my hypothesis that the pill bugs would eat more leaves.

Be sure you can support your conclusion using what you learned in the investigation.

Another Question

I wonder if pill bugs prefer moist or dry conditions. What would happen if I placed a pill bug between a dry area and a moist area?

Investigations often lead to new questions.

my SCIENCE notebook

Your Investigation

☐ **Explain and share your results.**

☐ **Tell what you conclude.**

☐ **Think of another question.**

How Scientists Work

Designing Investigations and Using Evidence

Scientists do investigations to answer questions about the natural world. They make and follow inquiry plans to gather, organize, and analyze information. The organized plans that scientists use to investigate questions are called scientific methods.

All investigations begin with a question or a problem that scientists want to solve. Scientists then make a plan that they will follow to answer the question.

Each of these scientists will follow his or her own plan.

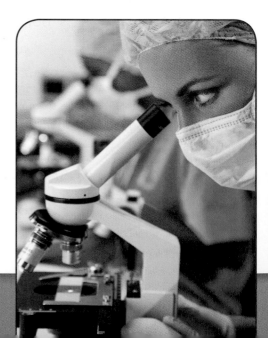

Scientists collect and record data. They analyze the data to explain what it shows. They make conclusions to answer their questions. Scientists use their data as evidence to support their explanations.

All investigations follow scientific methods, but there is no one set of steps that every investigation must follow. When planning an investigation, scientists follow the methods that will best help them to answer their question. Scientists have to be creative when they plan an investigation. A plan that helps to answer one question might not work as well to answer a different question.

Karl von Frisch

In the early 1900s, Austrian scientist Karl von Frisch wanted to find out whether honeybees could see color. He used scientific methods to investigate this question. His plan needed to be organized, but it also had to be creative.

In his investigation, von Frisch trained honeybees to feed from a dish that was set on a colored card. Then he placed the colored card in the middle of a group of gray cards. His hypothesis was that if most of the bees visited the colored card, then they could see color. During the investigation, most of the bees did visit the colored card. Von Frisch used this evidence to support his hypothesis. He concluded that honeybees see color.

Karl von Frisch won the Nobel Prize for his work with honeybees.

The honeybees were able to pick out the colored card from the gray cards.

SUMMARIZE

What Did You Find Out?

1 Why don't scientists use the same methods to investigate all scientific questions?

2 What evidence did Karl von Frisch use to support his hypothesis?

✋ Use Evidence to Answer a Question

Karl von Frisch also studied how bees find food. He observed that some worker bees left the hive to look for food. When these bees returned to the hive, they did a dance. Based on his observations, von Frisch concluded that the bees used different dances to communicate to other bees how far away the food was from the hive.

The graph shows data collected during an investigation of honeybee dances. Use the evidence in the graph to explain how the honeybee dance changes when food is farther away from the hive.

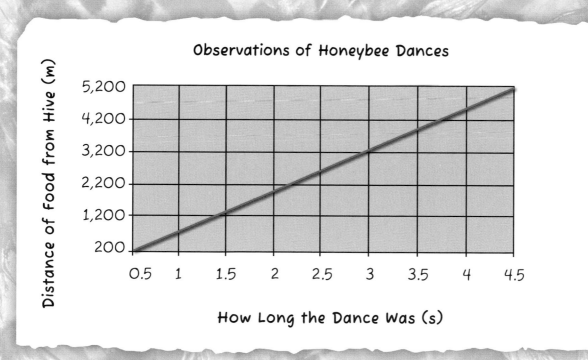

Science in a Snap!

Science in a Snap! Model Earth's Movements

Have a partner stand in the middle of a space. Your partner represents the sun. Hold a globe or ball, which will represent Earth. First, turn the globe or ball counterclockwise to **model** how Earth rotates. Now continue to turn the globe or ball as you carefully walk in a counterclockwise path around your partner to model how Earth revolves around the sun. In your model, when was it daytime on Earth? When was it nighttime? In your model, when did you complete one year of revolving around the sun?

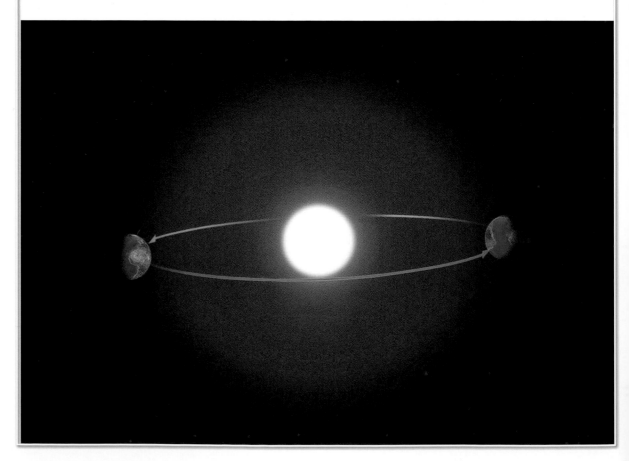

CHAPTER **2**

Science in a Snap! **Make a Model Metamorphic Rock**

Using different colors, roll clay into several pea-sized balls. The clay pieces are **models** of rocks. Place them close together on a piece of waxed paper. Place another piece of waxed paper over the top. Set a heavy book on top and press down on it. Remove the book and the waxed paper. You have made a model of a rock. **Observe** the clay. **Infer** how pressure changes rocks.

CHAPTER **3**

Science in a Snap! **Minerals in Your Classroom**

Look around your classroom. List as many objects as you can that are made from minerals. Remember that minerals can be combined to make new materials. **Share** your list with a partner. What are some objects you use that are made from minerals? How would your life change without the minerals that make these objects?

Investigate Star Positions

Question How do some stars appear to move across the sky?

Science Process Vocabulary

observe verb

Scientists often use tools to take a close look at, or **observe,** objects and events.

compare verb

When you **compare,** you tell how two or more things are alike or different.

I can compare these constellations.

Materials

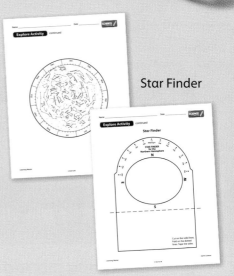

safety goggles

Star Finder

scissors

tape

What to Do

1 Put on your safety goggles. Cut out both parts of the Star Finder. Cut out the viewing window in the top piece of the Star Finder.

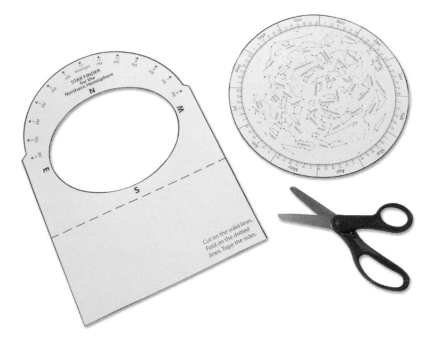

2 Fold the top piece of the Star Finder along the dotted lines. Tape the folds to form a pocket.

What to Do, continued

3 Slide the circle piece of the Star Finder into the pocket. Slowly turn the star wheel to the left to model time passing.

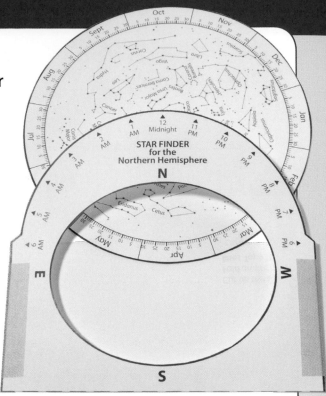

4 You can use a star finder to **observe** constellations, or patterns of stars, on a certain date. Set the Star Finder for July 21 at 9:00 PM. Record the season. Then observe the constellations you would see on that date. Record your observations in your science notebook. Now set the Star Finder for midnight. Describe any changes you observe in the appearance of the constellations.

5 Observe the constellations in a different season. Repeat step 4, but set the Star Finder for January 21 at 9:00 PM and midnight. Record your observations.

Record

Write and draw in your science notebook.
Use a table like this one.

Star Finder Observations

Date	Time	Season	Constellation Observations

Explain and Conclude

1. Did the star patterns change or stay the same as they moved across the sky?

2. On July 21, which constellations could you **observe** at midnight that you could not observe at 9:00 PM? Explain.

3. **Compare** the constellations you could observe on July 21 and January 21 at 9:00 PM. Tell what you **conclude** about observing stars in different seasons.

An early astrolabe

Investigate Moon Phases

Question How does the lighted part of the moon seem to change during the month?

Science Process Vocabulary

model noun

You can make and use a **model** to show how something in real life works.

observe verb

When you **observe,** you use your senses to learn about an object or event.

I can see the sun's light and feel its heat.

Materials

Moon Phase Pictures

Moon Calendar

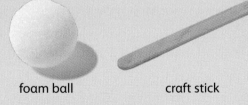

foam ball   craft stick

lamp

meterstick

What to Do

1 Use the Moon Calendar and Moon Phase Pictures to look for patterns in the moon's appearance. Follow the instructions on the Moon Phase Pictures.

2 Push the craft stick into the foam ball to form a handle. The foam ball is a **model** of the moon.

3 Place the lamp in a high place. The lamp represents the sun. Turn on the lamp.

4 Use a meterstick to **measure** a distance of 2 m from the lamp. Stand at that distance. Face the lamp and hold the foam ball in front of you. You represent the view from Earth.

What to Do,

5 Turn slowly to your left. Pause when you have made one quarter of a turn. The lamp should be directly to your right. **Observe** the foam ball. Draw your observations of the foam ball in your science notebook.

6 Continue turning to your left. Pause when you have completed another quarter turn (one-half turn total). The lamp should be directly behind you. Draw your observations of the foam ball.

7 Continue turning to your left. Pause when you have completed another quarter turn (three-quarter turn total). The lamp should be directly on your left. Observe the foam ball and draw your observations.

8 Complete your turn so you are facing the lamp again. Draw your observations of the foam ball.

Record

Write and draw in your science notebook.
Use a table like this one.

Observations of Model Moon

One-Quarter Turn	One-Half Turn	Three-Quarter Turn	One Complete Turn

Explain and Conclude

1. How did the lighted part of the surface of the **model** moon change each time you turned?

2. Based on your **observations,** work with a partner to come up with an explanation for the cause of moon phases.

Think of Another Question

What else would you like to find out about how the lighted part of the moon seems to change? How could you find an answer to this new question?

Earth's moon

Think
Like a Scientist

How
Scientists Work

Using Data to Separate Fact and Opinion

If someone says a diamond is the hardest mineral, that is a fact. A fact is a statement that is based on evidence. You can do scientific investigations to show that a diamond is the hardest mineral.

However, if someone says that a diamond is the prettiest mineral, they are giving an opinion. An opinion is what someone thinks. An opinion is not based only on evidence. An opinion cannot be shown to be true by doing scientific investigations.

Facts are supported by evidence that is based on observations and data. As you read, you evaluate which statements are supported by evidence and which are not. You use data and what you already know to analyze the statements. Separating fact from opinion helps you make choices about how you will use the information you have read.

Using Facts to Answer Research Questions

Scientists often read about the work of other scientists to help answer their research questions. Scientists evaluate information just as you do when you read an article in a newspaper or magazine. They separate facts from opinions. They look to see which statements are based on evidence. If information is not supported by evidence, scientists will not use it to answer their questions.

Using Facts to Share Research Results

When scientists share the results of an investigation, they use their observations and the data from their experiments to explain their results. If scientists use their results to draw a conclusion, they must share what evidence they used. Other scientists will not accept conclusions based on opinions rather than facts.

Evaluating Information When scientists look for information, they can ask themselves questions about what they read to decide whether the information is a fact or an opinion.

- Are the statements supported by evidence, or do they give the author's opinion?

- What evidence does the author use to support his or her statements?

- Does this information match what I already know or have found through my own research?

- Do other sources agree or disagree with this information?

Using data to separate fact from opinion helps scientists make sure that they draw accurate conclusions about the natural world.

SUMMARIZE

What Did You Find Out?

1 How is a fact different from an opinion?

2 Why is it important for scientists to separate fact from opinion when doing research?

Identify Fact and Opinion

You can use the same questions that scientists use to evaluate what you read. Find an article about a science topic in a magazine or a newspaper, or print one from the Internet. Complete the following steps to identify statements of fact and opinion in the article.

- First, underline the facts in the article. What evidence is included that tells you each statement is a fact?

- Next, circle any opinions you see. How can you tell these statements are opinions?

- Finally, summarize the facts and opinions in your article to the class. Would you use this article to answer a research question? Why or why not?

Math in Science

Using Tables to Identify Objects

Scientists often organize their data in tables. In a table, facts and figures are arranged in columns and rows. Displaying data in a table is one way that scientists can share the results of an investigation.

A table is a good way to organize numerical data.

Rock Particle Sizes

Name	Kind	Particle Size (mm)
Mud	Clay	smaller than 1/256
	Silt	1/256 - 1/16
Sand	Sand	1/16 - 2
Gravel	Granule	2 - 4
	Pebble	4 - 64
	Cobble	64 - 256
	Boulder	larger than 256

Parts of a Table Tables organize information so that it is easy to find. The title of a table shows what data are displayed. The title of the table shown below is *Colors of Rocks*. By looking at the title, you know that the table contains information about rocks and their colors.

Tables usually organize data in boxes, or cells. Cells can contain words, numbers, or drawings. The cells are organized in rows and columns. Rows of cells are read from left to right, and columns are read from top to bottom.

Headings in the table show what information is included in each column or row. The heading of the first column of the table below is *Rock Name,* so you know that the cells in this column contain the names of rocks. The heading of the second column is *Color.* The cells in this column contain information about each rock's color.

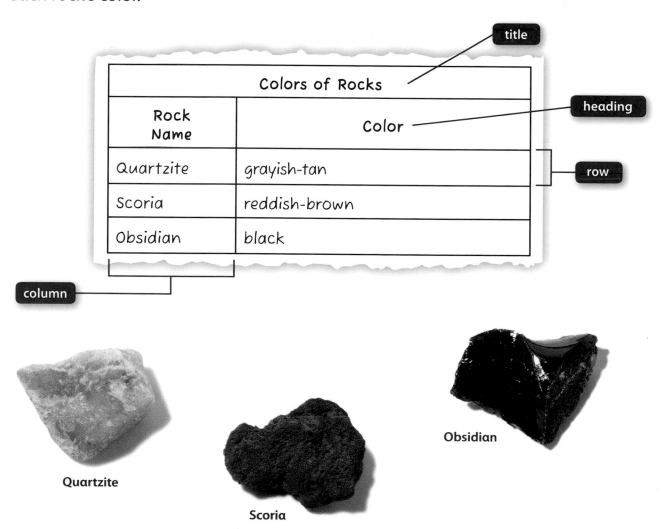

title

heading

row

Colors of Rocks	
Rock Name	Color
Quartzite	grayish-tan
Scoria	reddish-brown
Obsidian	black

column

Quartzite

Scoria

Obsidian

Using Tables Tables often show data about different objects. You have observed a table of rocks of different colors. Rocks are made of minerals. The table below organizes facts about the properties of three different minerals. You can look at the properties of each mineral to compare the minerals. You could also use this table to help you identify an unknown mineral. If the properties of the unknown mineral match the properties of one of the minerals listed in the table, you can use that information to identify the unknown mineral.

Mineral Properties

Mineral	Color	Luster (How a Mineral Reflects Light)
Rose quartz	pink	nonmetallic (glassy)
Hornblende	greenish-gray	nonmetallic (glassy to dull)
Pyrite	gold	metallic

SUMMARIZE

What Did You Find Out?

1. Which parts of a table help you find out what data are being displayed?

2. How are the data in a column arranged?

3. How could you use a table of mineral properties to identify an unknown mineral?

🖐 Practice Using Tables to Identify Minerals

Look at the photos of minerals below. Use the Mineral Properties table on page 100 to identify each mineral.

Mineral C

Mineral A

Mineral B

1 Observe the properties of each mineral above.

2 Compare the mineral's properties to the properties listed in the table.

3 Find the mineral in the table that shares the same properties.

4 Identify each mineral. Explain how you made your identification.

Investigate Minerals

Question How can you identify minerals by their properties?

Science Process Vocabulary

compare verb

When you **compare,** you tell how objects or events are alike and different.

observe verb

When you **observe,** you use your senses to learn about an object or event.

The hand lens will help me observe the luster of the minerals.

Materials

safety goggles

mineral samples

hand lens

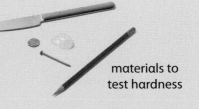
streak plate

materials to test hardness

Mineral Identification Key

What to Do

1 Put on your safety goggles. Choose a mineral that you want to identify. Record its number in your science notebook.

2 **Observe** the sample. Record its color and luster in your science notebook. Luster is how a mineral reflects light. Luster can be metallic or nonmetallic. You can see examples of each type of luster in the photos below.

metallic luster

nonmetallic luster (glassy)

nonmetallic luster (dull)

3 Put the streak plate flat on the table. Rub the mineral on it. Look at the powder on the plate. What color is the powder? Record your observations. The color of a mineral's powder is its streak.

What to Do, continued

4 Hardness is recorded as a Mohs number. An object with a low Mohs number can be scratched by an object with a higher Mohs number. Use the items on the Hardness of Materials chart and other materials to test the hardness of minerals. Record the Mohs hardness.

Hardness of Materials

Material	Mohs Hardness Number
pencil tip	1
fingernail	2.5
penny	3.5
iron nail	4.5
butter knife	5.5
quartz	7

5 Repeat steps 1–4 for the other minerals.

6 Use the Mineral Identification Key to identify each of the minerals you tested.

Record

Write in your science notebook.
Use a table like this one.

Mineral Identification

Property	Mineral Number				
	_____	_____	_____	_____	_____
Color					
Luster					

Explain and Conclude

1. **Compare** the minerals. How are they the same? How are they different?

2. How did you identify each mineral you **observed?**

3. Which properties were most helpful in identifying the minerals?

Think of Another Question

What else would you like to find out about identifying minerals by their properties? How could you find an answer to this new question?

Goat's Paradise,
Rumbling Falls Cave System,
Spencer, Tennessee

Gypsum formation

Investigate Rocks

Question How can you use properties to identify igneous, sedimentary, and metamorphic rocks?

Science Process Vocabulary

classify verb

When you **classify,** you put things into groups according to their characteristics.

I can see and feel rocks to classify them.

share verb

I learned that the rock I observed was sandstone.

When you **share** results, you tell or show what you have learned.

Materials

3 index cards

rock group 1   rock group 2

rock group 3

hand lens   Rock Properties chart

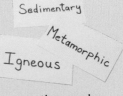

Sedimentary
Metamorphic
Igneous

sorting cards

What to Do

1 Label the index cards **1, 2,** and **3** to represent each of the 3 rock groups. Choose 1 rock from each group. Place the rocks on the corresponding index cards.

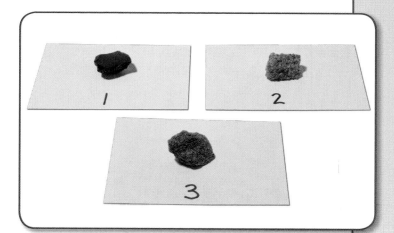

2 Use the hand lens to **observe** the properties of each rock. Record your observations in your science notebook.

What to Do, continued

3 Use your observations and the Rock Properties chart to find the name of each rock. Record the name.

4 Use the Rock Properties chart to **classify** each rock as igneous, metamorphic, or sedimentary. Record your classification.

5 **Share** your rocks and cards with others. Place each rock under the correct sorting card your teacher has supplied: Igneous, Metamorphic, or Sedimentary. Observe the rocks in each group. Talk about the properties that rocks in a group share.

Sedimentary

Metamorphic

Igneous

Record

Write in your science notebook.
Use a table like this one.

Classifying Rocks

Rock	Properties	Rock Name	Rock Classification
1			
2			
3			

Explain and Conclude

1. Explain how you used properties to **classify** each rock.

2. What were the names of the rocks in each group?

3. How were the rocks in each group alike?

Think of Another Question

What else would you like to find out about how to use properties to identify and classify rocks? How could you find an answer to this new question?

These basalt pebbles are igneous rocks.

Investigate Soil Properties

Question Is loamy soil, clay soil, or sandy soil best for growing corn plants?

Science Process Vocabulary

predict verb

You **predict** when you say what you think will happen.

> I predict that the plants will grow well in the loamy soil.

measure verb

You **measure** when you use tools to find the size, length, or amount of something.

> I can use a ruler to measure the height of the plant.

Materials

safety goggles

sandy soil

clay soil

loamy soil

hand lens

3 corn seedlings

ruler

water

spoon

What to Do

1 Put on your safety goggles. **Observe** the physical properties of each type of soil. Describe its color and odor. Which soil has the most pieces of decaying materials? Which has more humus? Record your observations of these physical properties in your science notebook.

2 Use a hand lens to observe the kinds and sizes of the particles in each soil. Record your observations.

3 **Predict** which soil will be best for growing a corn seedling. Record your prediction.

What to Do, continued

4 Use the spoon to carefully plant one corn seedling in each cup. Make sure the roots are covered with soil. The stem should be above the surface of the soil.

5 Use a metric ruler to **measure** the height of each seedling. Record your measurements.

6 Place the seedlings in a sunny place. Give each seedling 5 spoonfuls of water every day. After 1 week, measure the seedlings again. Record your measurements.

Record

Write and draw in your science notebook.
Use tables like these.

Soil Observations

Soil Type	Physical Properties	Particles (Kinds and Size)
Sandy		

Growth of Corn Seedlings

Soil Type	Beginning Height of Seedling (cm)	Height of Seedling After One Week (cm)
Sandy		

Explain and Conclude

1. Did your results support your **prediction?** Explain.

2. What **observations** and information did you use to make your prediction?

3. What can you **conclude** about which kind of soil is best for growing corn seedlings? Use your observations as evidence.

Think of Another Question

What else would you like to find out about growing corn plants in different kinds of soil? How could you find an answer to this new question?

Investigate Soil and Water

Question How can you find out whether some soils absorb more water than others?

Science Process Vocabulary

variable noun

A **variable** is something in an experiment that you can change.

You change only one variable in an experiment. You keep all other variables the same.

I will change the type of soil I test. I will keep the other variables the same.

Materials

safety goggles

soil A

soil B

soil C

hand lens

2 plastic foam cups

spoon

tape

2 plastic cups

graduated cylinder

water

stopwatch

Do an Experiment

Write your plan in your science notebook.

Make a Hypothesis

In this investigation, you will choose two kinds of soil: loamy, clay, or sandy. You will pour water into two cups with different kinds of soil. The cups have holes in the bottom. You will measure the amount of water that drains through the soils into the cups. How will the amount of water that drains through each cup compare? Will it be the same or different?
Write your **hypothesis.**

Identify, Manipulate, and Control Variables

Which variable will you change?
Which variable will you observe or measure?
Which variables will you keep the same?

What to Do

1 Put on your safety goggles. Select 2 soil samples to test. Use the hand lens to **observe** the properties of each type of soil, including its color, texture, and particle size. **Classify** each soil as sandy, clay, or loamy. Record your observations and classifications in your science notebook.

2 Use a pencil to poke 5 small holes in the bottom of each plastic foam cup.

3 Fill 1 cup with one of the soil types you selected. Fill the other cup with the second type of soil. Label the soil type on each cup.

4 Place each cup of soil inside a plastic cup. **Measure** 75 mL of water with the graduated cylinder. Pour the water into one cup of soil. Repeat for the second cup of soil.

5 After 10 minutes, remove the cups of soil from the plastic cups. Use the graduated cylinder to measure how much water is in the bottom of each plastic cup. Record your measurements.

6 To find out how much water each type of soil absorbed, subtract the amount of water in the plastic cup from 75 mL. Record your **data.**

Record

Write in your science notebook.
Use a table like this one.

Soil Observations

Soil Properties	Soil Type	Amount of Water Poured on Soil (mL)	Amount of Water in Cup (mL)	Amount of Water Absorbed (mL)
		75 mL		
		75 mL		

Explain and Conclude

1. Did the results support your **hypothesis?** Explain.

2. **Compare** the amount of water absorbed by each kind of soil. What can you **conclude** about different kinds of soil and the amount of water they absorb?

3. **Share** your results with the results of others. Explain any differences.

Think of Another Question

What else would you like to find out about whether some soils absorb more water than others? How could you find an answer to this new question?

Investigate Weathering

Question How can you model weathering and erosion?

Science Process Vocabulary

model noun

A **model** can show how a process, such as weathering, works.

predict verb

When you want to explain what you think will happen, you **predict**.

I predict there will be no change when sandstone is placed in water.

Materials

safety goggles

sandstone

paper towel

water

graduated cylinder

2 jars with lids

hand lens

stopwatch

vinegar

limestone

What to Do

1 Put on your safety goggles. **Predict** what will happen if you rub 2 pieces of sandstone together. Hold the pieces of sandstone over the paper towel and rub them together for a few seconds. **Observe** what happens. Record your observations in your science notebook.

2 **Measure** 250 mL of water into a jar. Place 5 pieces of sandstone in the jar. Put the lid on the jar securely.

3 Use the hand lens to observe the sandstone and the bottom of the jar. Record your observations.

What to Do, continued

4 Predict what will happen if you shake the jar. Then shake the jar for 3 minutes. Use the stopwatch to time how long you shake the jar. This is a **model** of physical weathering and erosion. Observe the sandstone and the bottom of the jar. Record your observations.

5 Predict what will happen if limestone is placed in vinegar. Measure 100 mL of vinegar into the other jar. Place 5 pieces of limestone in the jar. Put the lid on the jar securely. This is a model of chemical weathering.

6 Use a hand lens to observe the limestone in vinegar. Record your observations every day for 3 days.

Record

Write in your science notebook. Use tables like these.

Physical Weathering and Erosion

	Rubbing Sandstone Together	Shaking Sandstone in Water
Prediction		
Observation		

Chemical Weathering

	Soaking Limestone in Vinegar
Prediction	
Observation: Day 1	

Explain and Conclude

1. Do the results support your **predictions?** Explain.

2. Explain how one of the **models** shows physical weathering. Explain how one of the models shows chemical weathering.

3. Explain how one of the models shows erosion.

Think of Another Question

What else would you like to find out about weathering and erosion? How could you find an answer to this new question?

Pictured Rocks, Lake Superior, Michigan

Investigate Erosion

Question How does the way water moves on soil affect the way the soil moves?

Science Process Vocabulary

variable noun

A **variable** is a part of an experiment that you can change.

You change only one variable while you keep all the other parts the same. You control the parts that do not change.

I will only change 1 variable in the experiment.

Materials

safety goggles   3 plastic containers

tape

small paper cup

soil

3 wood blocks

plastic foam cup

paper clip

water

graduated cylinder

ruler

Do an Experiment

Write your plan in your science notebook.

Make a Hypothesis

In this investigation, you will pour water through large and small holes in cups onto soil. Water moves slowly through small holes and quickly through large holes. How will hole size affect the amount of soil erosion you observe?
Write your **hypothesis.**

Identify, Manipulate, and Control Variables

Which variable will you change?
Which variable will you observe or measure?
Which variables will you keep the same?

What to Do

1. Put on your safety goggles. Label the plastic containers **1, 2,** and **3.** Put 1 paper cupful of soil at one end of each of the containers. Put a wood block under the same end of the container as the soil. You will not pour any water into container 3.

2. Use the paper clip to poke 2 small holes in the bottom of the plastic foam cup.

What to Do, continued

3 **Measure** 75 mL of water using the graduated cylinder. Decide how high to hold the plastic foam cup above container 1. Record the height you choose. Pour the water from the graduated cylinder into the plastic foam cup. Hold the cup over the soil until all the water has dripped onto the soil.

4 Use the paper clip to make the holes in the plastic foam cup larger. Measure 75 mL of water using the graduated cylinder. Hold the plastic foam cup above container 2 at the same height as in step 3. Pour the water into the plastic foam cup.

5 **Observe** the soil in each container. **Compare** container 3 to containers 1 and 2. Record how much soil was eroded in each container. **Estimate** how much of the cupful of soil was moved to the other end of the container.

Record

Write in your science notebook.
Use a table like this one.

Water Movement and Soil Erosion

Container	Height of Foam Cup Above Container (cm)	Size of Holes (small or large)	Amount of Erosion
1			
2			
3	No water poured		

Explain and Conclude

1. Do the results support your **hypothesis?** Explain.

2. Which container with water had the least amount of erosion? Which container with water had the most erosion?

3. What can you **conclude** about how fast water is poured onto soil and the amount of erosion that occurs?

Think of Another Question

What else would you like to find out about how water affects soil erosion? How could you find an answer to this new question?

Amazon River Basin, Brazil

Investigate Earthquakes

Question How can you use a model to understand one way in which Earth's plates can move?

Science Process Vocabulary

model noun

You can use a **model** to learn about events in the natural world that you can't study up close.

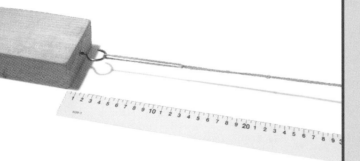

measure verb

When you **measure,** you find out how much or how many.

I can use a meterstick to measure how far the block moves. I can use this model to describe what happens with Earth's plates.

Materials

safety goggles

3 rubber bands

wooden block with hook

wooden block without hook

tape

meterstick

What to Do

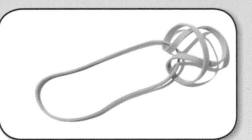

1 Put on your safety goggles. Loop the rubber bands together to form a chain, as shown.

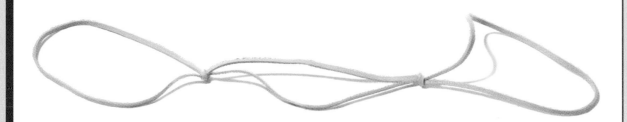

2 Place the wooden block with the hook on a desk. Place the other block on top of the first block. Attach one end of the rubber band chain to the hook on the bottom block.

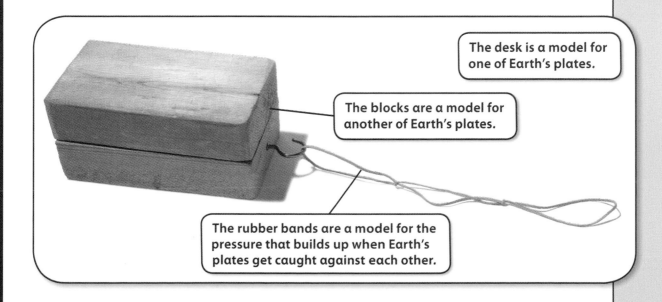

The desk is a model for one of Earth's plates.

The blocks are a model for another of Earth's plates.

The rubber bands are a model for the pressure that builds up when Earth's plates get caught against each other.

What to Do, continued

3 Place tape on the desk to mark the starting position of the blocks. Place the meterstick on the edge of the tape.

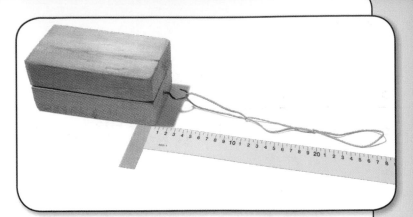

4 **Predict** what will happen when you pull on the rubber bands. Record your prediction in your science notebook.

5 Slowly pull the rubber bands back 1 cm at a time. **Observe** how far the rubber bands are stretched as you pull. Also observe the blocks. Stop pulling when the blocks slip and move forward quickly. Record how many centimeters the rubber bands stretched before the blocks moved. **Measure** and record the total distance that the blocks moved.

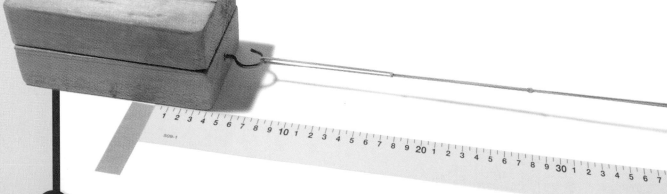

6 Repeat step 5 two more times. Record your observations.

Record

Write in your science notebook.
Use a table like this one.

Observations of Blocks and Rubber Bands

	How Far the Rubber Bands Were Stretched (cm)	How Far the Blocks Moved (cm)
Trial 1		
Trial 2		

Explain and Conclude

1. Did your results support your **prediction?** Explain.

2. What happened to the tension of the rubber bands as you pulled them farther? How did the tension change when the blocks moved?

3. How did this **model** help you understand one way in which Earth's plates can move when pressure builds between them?

Think of Another Question

What else would you like to find out about the movement of Earth's plates in an earthquake? How could you find an answer to this new question?

Port Island, Kobe, Honshu, Japan

Earthquakes can cause serious damage to towns and cities.

Investigate Mountain Building

 **Question** How can you use clay to investigate how folded mountains form?

Science Process Vocabulary

Materials

model noun

You can make and use a **model** to represent an object or process.

You can use a model to learn about processes that change Earth's surface.

I will use a model to show how some mountains form.

3 different colors of clay

ruler

130

What to Do

1 Press the clay into three flat ovals. Each oval should be about 12 cm long and 7 cm wide. Use a ruler to **measure** your ovals.

2 Stack the clay ovals on top of each other to form 3 layers. **Observe** the layers from the side. Draw your observations in your science notebook.

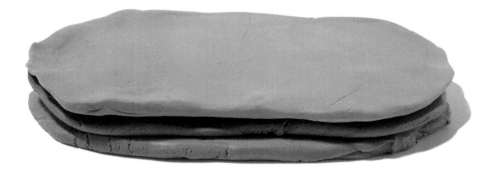

What to Do, continued

3 **Predict** what will happen to the layers when you gently push the ends of the clay toward each other. Record your predictions.

4 Gently push the ends of the clay toward each other. Record your observations.

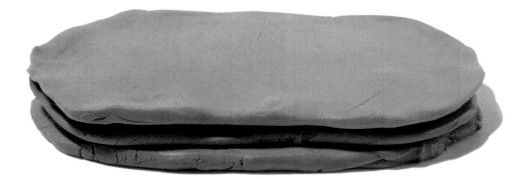

5 Predict what will happen if you change the force that you use to push the layers. Record your predictions. Do 2 more trials with different amounts of force. Record your predictions and observations.

6 How can you push the clay so that it makes both upward and downward folds? Try it! Record your observations.

Record

Write and draw in your science notebook.
Use a table like this one.

Effects of Pushes on Clay

Force of Push	Prediction	Draw Your Observations
No push		
Gentle		

Explain and Conclude

1. How did the amount of force affect the way the clay folded? Did these results support your **predictions?**

2. How can you use your **observations** and **conclusions** from this **investigation** to explain how some mountains form? Tell how the layers of clay and your pushes are a model of the process of mountain building.

3. Use the results of the investigation to **infer** how folded mountains form.

Think of Another Question

What else would you like to find out about how to use clay to investigate folded mountain formation? How could you find an answer to this new question?

Mount Rundle,
Banff National Park,
Alberta, Canada

Investigate the Water Cycle

Question What happens to a model snow-covered mountain when sunlight shines on it?

Science Process Vocabulary

model noun

You can make and use a **model** to learn how a process such as the water cycle works.

predict verb

You **predict** when you use your observations and prior knowledge to say what will happen.

I predict that cold temperatures will cause the water in the pond to freeze.

Materials

safety goggles

soil

plastic bin

crushed ice

plastic wrap

tape

wooden block

134

What to Do

1 Put on your safety goggles. Add the loamy soil to one end of the plastic bin. Shape the soil into a mound. Place the crushed ice on top of the mound of soil. The soil and ice form a **model** of a snow-covered mountain.

2 Cover the bin with plastic wrap. Tape the plastic wrap to the bin. The bin is a model of the mountain and the air around it. Set the bin in a sunny spot and place a block under one end.

What to Do, continued

3 **Observe** and draw your model. Identify where water is located in your model and whether the water is a solid, liquid, or gas. Label the water in your drawing. **Predict** what will happen to the ice after 1 hour. Record your predictions.

4 Observe your model after 1 hour. Has the water changed? If so, how? Record your observations and draw your model. Predict what will happen to the water after 24 hours. Record your predictions.

5 Observe your model after 24 hours. Tap the plastic wrap and observe what happens. Has the water changed? If so, how? Draw your model and record your observations.

Record

Write and draw in your science notebook.
Use a table like this one.

Model Mountain and Air Around It

Time	Prediction	Observations	Drawing
Start			
After 1 hour			

Explain and Conclude

1. Did your results support your **predictions?** Explain.

2. Describe how water changed throughout the **investigation.** What caused these changes?

3. Use what you **observed** about your **model** to explain what happens in the water cycle.

Think of Another Question

What else would you like to find out about how you can use a model to learn about the water cycle?
How could you find an answer to this new question?

Rain falling to Earth is one step in the water cycle.

Investigate Weather

Question How can you use weather patterns to predict weather?

Science Process Vocabulary

measure verb

Scientists use tools to accurately **measure** weather conditions.

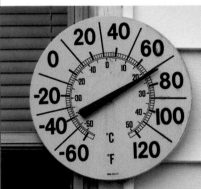

I can use a thermometer to measure air temperature.

analyze verb

When you **analyze** data, you look for patterns, or things that happen over and over.

The weather pattern I observed was that if dark clouds formed, rain followed.

Materials

weather station

barometer

hygrometer

large calendar

138

What to Do

1 The class will **measure** weather conditions outside the school for 2 weeks. Each group will track one condition: precipitation, cloud cover, temperature, wind speed and direction, air pressure, or humidity. Decide with your group which weather condition you want to measure. Select the appropriate weather tool. If you are monitoring cloud cover, you will not use a weather tool.

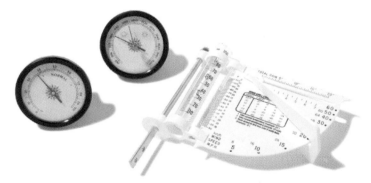

2 Set up your weather tool outside. Make sure the tool is not blocked by buildings, trees, or other objects that would interfere with a proper measurement. Measure your weather condition. Recheck your measurement to make sure it is correct. Record the **data** in your science notebook.

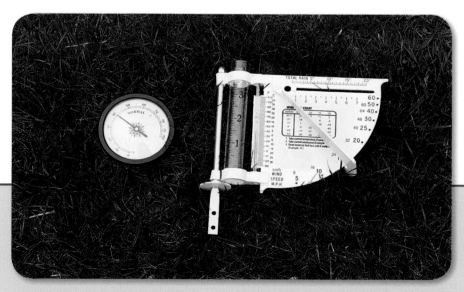

What to Do, continued

3 Measure and record your weather condition each day for 2 weeks. Record the time of day you collected the weather data. Take your measurements at the same time each day.

4 Compile the data from all groups onto a class calendar. Look for patterns in the data. Use the patterns to **predict** the next day's weather. Then **compare** your prediction to the actual weather.

October

Sunday	Monday	Tuesday	Wednesday	Thursday	Friday	Saturday
				1 — 5mm of rain partly cloudy 70°C 13 mph from East 1000 mb 87% humidity	2 — 14mm of rain partly cloudy 77°C 15 mph from Southwest 1053 mb 92% humidity	3
4	5 — 52mm of rain very cloudy 75°C 17 mph from North 1002 mb 96% humidity	6	7	8	9	10
11	12	13	14	15	16	17
18	19	20	21	22	23	24
25	26	27	28	29	30	31

Record

Write in your science notebook.
Use a table like this one.

| Weather Data: Week 1
 Weather Condition I Will Observe: _____ | | |
Date	Time	Observations/Data

Explain and Conclude

1. How did the **data** for the weather condition you **measured** vary during the 2 weeks?

2. What patterns were you able to see when you **analyzed** the class data?

3. Was your **prediction** of the next day's weather supported? What information did you use to make your prediction?

Think of Another Question

What else would you like to find out about how you can use weather patterns to predict weather? How could you find an answer to this new question?

Priest Lake, Idaho

What weather conditions can you observe?

Do Your Own Investigation

Question **Choose one of these questions, or make up one of your own to do your investigation.**

- How can you use shadows caused by the sun to tell time?
- How does the rate of cooling affect how alum crystals form?
- What happens when pure water and tap water evaporate?
- What happens to sand particles of different sizes when they are blown by the wind?
- How does gravity affect soil on a slope?
- How is air temperature different over land and water?

Science Process Vocabulary

hypothesis noun

When you make a **hypothesis,** you state a possible answer to a question that can be tested by an experiment.

If I place an alum solution in the refrigerator, the crystals will form more quickly.

Open Inquiry Checklist

Here is a checklist you can use when you investigate.

- ☐ Choose a **question** or make up one of your own.

- ☐ Gather the materials you will use.

- ☐ If needed, make a **hypothesis** or a **prediction.**

- ☐ If needed, identify, manipulate, and control **variables.**

- ☐ Make a **plan** for your **investigation.**

- ☐ Carry out your plan.

- ☐ Collect and record **data. Analyze** your data.

- ☐ Explain and **share** your results.

- ☐ Tell what you **conclude.**

- ☐ Think of another question.

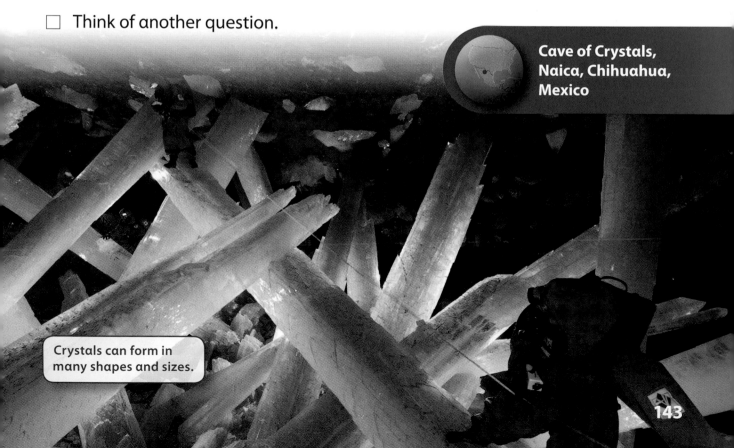

Cave of Crystals,
Naica, Chihuahua,
Mexico

Crystals can form in many shapes and sizes.

Write About an Investigation

Crystal Formation

The following pages show how one student, Jason, wrote about an investigation. As he read that minerals make up rocks, Jason wondered whether crystals form at different rates at different temperatures. He decided to do an investigation. Here is what he thought about to get started:

Alum Crystal

- Jason wanted to use simple materials to make mineral crystals. He decided to make an alum solution.

- He decided to test whether the temperature at which the crystals formed affected how fast the crystals formed.

- He would keep two alum solutions in places with different temperatures to observe in which the crystals formed first.

Model

Question

How does the rate of cooling affect how crystals form?

> Make sure your question states clearly what you want to find out.

Materials

alum	2 plastic cups
water	2 craft sticks
container (larger than 250 mL)	2 strings (9 cm each)
plastic spoon	rubber band

> Be sure to list everything you will need.

Your Investigation

Now it's your turn to do your investigation and write about it. Write about the following checklist items in your science notebook.

☐ **Choose a question or make up one of your own.**

☐ **Gather the materials you will use.**

Model

My Hypothesis

If I keep one alum solution at room temperature and place one alum solution in the refrigerator, then crystals will form more quickly in the solution in the refrigerator.

Be specific about what you think the results of the experiment will be.

 **Your Investigation**

☐ **If needed, make a hypothesis or prediction.**

Write your hypothesis or prediction in your science notebook.

Model

Variable I Will Change

One cup with alum solution will be kept at room temperature. The other cup with alum solution will be placed in the refrigerator.

Variable I Will Observe or Measure

I will observe how quickly crystals form.

Variables I Will Keep the Same

Everything else will be the same. Both solutions will contain the same amount of alum and water. Both solutions will be in the same kind of cup. I will use the same kind of string in each cup. I will observe the crystals for the same amount of time.

Answer these three questions:
1. What one thing will I change?
2. What will I observe or measure?
3. What things will I keep the same?

Your Investigation

☐ **If needed, identify, manipulate, and control variables.**

Write about the variables for your investigation.

Model

My Plan

1. Pour 250 mL of very warm water into a container.

2. Stir 1 spoonful of alum at a time into the solution. Stop adding alum when the alum no longer dissolves in the solution.

3. Pour 100 mL of the solution into each cup. Be careful not to pour any undissolved crystals into the cups.

4. Tie a string around the middle of each craft stick. Place each craft stick across the top of a cup. The string should not touch the bottom of the cup.

5. Place 1 cup in the refrigerator. Place the other cup in a room-temperature place.

6. Observe the cups every day for 3 days. Look carefully at the string.

> Your steps should give details about everything you will do.

SCIENCE my science notebook ✏ **Your Investigation**

☐ **Make a plan for your investigation.**

Write the steps for your plan.

Model

I carried out all 6 steps of my plan.

Make notes about any changes you made to the plan. Jason didn't make any changes to his plan.

 **Your Investigation**

☐ **Carry out your plan.**

Be sure to follow your plan carefully.

Model

Observations of Crystals in Both Cups

Day	Cup in Refrigerator	Cup at Room Temperature
1	No crystals.	No crystals.
2	Very small crystals are around the string. They are too small to measure.	No crystals.
3	The crystals continue to get larger.	Very small crystals have formed around the string.

My Analysis

Crystals formed in both of the alum solutions, but crystals formed more quickly in the refrigerator.

 **Your Investigation**

☐ **Collect and record data. Analyze your data.**

Collect and record your data, and then write your analysis.

Examine your data carefully to understand what happened.

How I Shared My Results

First I showed real life examples of minerals with visible crystals to the class. Next I explained the steps I took to carry out my plan. Finally, I shared the results from my investigation.

My Conclusion

Crystals formed as each solution cooled. Crystals formed more quickly in the refrigerator because the solution cooled more quickly. The results support my hypothesis.

Your results will not always support your hypothesis. Jason decided to repeat his investigation to look for possible errors in his data.

Another Question

How does the rate of evaporation of an alum solution affect the size of crystals that form?

Investigations often lead to new questions for inquiry.

SCIENCE my notebook

Your Investigation

- [ ] **Explain and share your results.**
- [ ] **Tell what you conclude.**
- [ ] **Think of another question.**

How
Scientists Work

glacier

Using Models

Scientists use models to better understand the natural world. Models can show how something in real life looks or how it works.

2-Dimensional Models An illustration is a 2-dimensional model. It might show features or processes, such as the form of a glacier and how the glacier can change the land. Drawings, diagrams, and maps are other examples of 2-dimensional models.

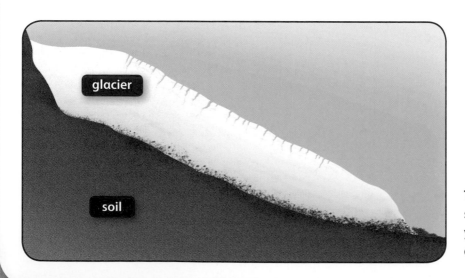
glacier

soil

This 2-dimensional model shows what you would see if you could cut down through a glacier into the soil.

Computer Models Scientists use computer models to show how things work and change in the natural world. Computer models can also be used to make and test predictions.

The Florida map is a computer model that combined different types of data to produce a single picture. The map shows when scientists predict that a common weed will begin to grow in different parts of Florida. Scientists used data about soil and air temperatures to make the computer model.

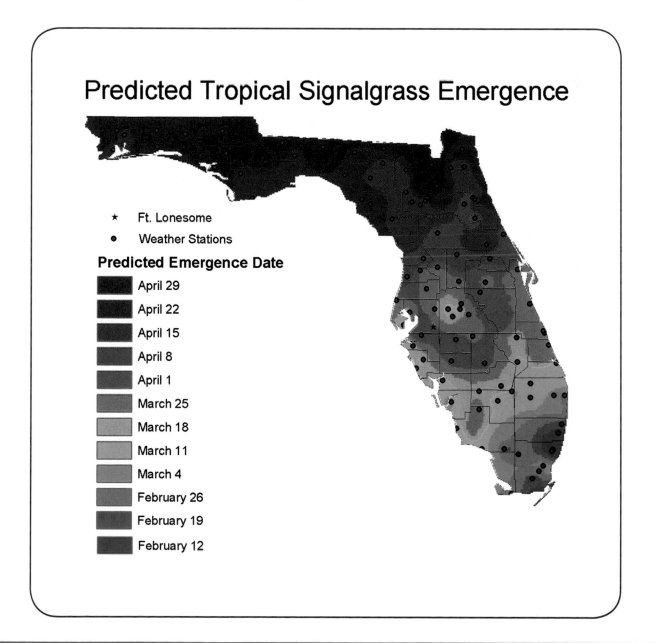

Predicted Tropical Signalgrass Emergence

★ Ft. Lonesome
● Weather Stations

Predicted Emergence Date
- April 29
- April 22
- April 15
- April 8
- April 1
- March 25
- March 18
- March 11
- March 4
- February 26
- February 19
- February 12

3-Dimensional Models A 3-dimensional model can be made from different materials to represent a real thing. Scientists can use 3-dimensional models to show a process or test predictions. Changes to Earth's surface can be demonstrated using models like the hills in the photo. Different kinds of plants could be planted on the model hills to see if the plants can help prevent erosion. The models could be made with different kinds of soil to see how water moves and deposits the soils.

SUMMARIZE
What Did You Find Out?

1 What are 3 ways scientists use models?

2 What are some different kinds of models?

Model Wind Erosion

You can make a model to study how wind erodes sand.

1. Put on your safety goggles. Form a small pile of sand into a hill on a large sheet of paper. Record the shape of the pile.

2. Gently blow through the straw onto the hill. Observe the hill. Record how the shape of the hill changes.

3. Repeat step 2 several more times. Each time move the straw around and blow onto the hill in a different place or from a different direction.

4. Tell what you learned about wind erosion from your model.

Science in a Snap!

Science in a Snap! Estimate Volume

Observe the size of a small paper cup. **Estimate** the number of centimeter cubes that will fit in the cup. Record your estimate in your science notebook. **Count** the number of cubes it takes to fill the cup. Was your estimate close to the number of cubes that fit in the cup? Repeat the steps with a tall plastic cup and a short plastic cup. **Compare** your results with others in your class.

CHAPTER
2

Science in a Snap! Observe Changes in Ice

Make a small container out of foil. Place an ice cube in the container. After 1 hour, **observe** what has happened to the cube. How has the shape of the ice cube changed? Explain how you could change the liquid water back into solid ice. What are some common ways you use both liquid and solid water?

CHAPTER
3

Science in a Snap! Changing Motion

Hold a rubber ball in front of you, higher than your head, and drop it. **Observe** the motion of the ball. Describe how the ball's direction and speed change when it hits the ground. What force is acting on the ball when it is falling? What force causes the ball's motion to change?

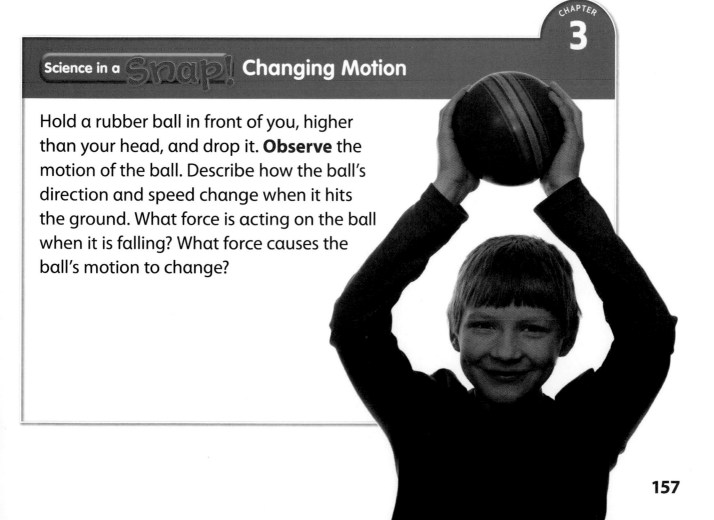

Science in a Snap! Compare Magnetic Attraction

Place a pencil in a ball of clay so that the pencil stands straight up. Place a donut magnet around the pencil so that it rests on the clay. **Predict** what will happen if you place another magnet around the pencil. **Observe** what happens. Remove the top magnet. Flip it over, and place it around the pencil again. Observe what happens to the magnets. Why do you think the magnets acted differently?

Science in a Snap! Observe Changes in Sand

Half-fill a plastic jar with sand. Touch the sand and **observe** its temperature. Does the sand feel warm or cool? Place the lid tightly on the jar. Take turns with a partner to shake the jar for 5 minutes. Remove the lid and touch the sand again. Does the sand feel warmer or cooler? What type of energy caused the temperature of the sand to change?

Science in a Snap! **Make Sounds**

Cut a string about 80 cm long. Tie a metal spoon to the middle of the string. Wrap the ends of the string around your two index fingers. Then use your index fingers to hold your ears closed. At the same time, lean forward and allow the spoon to hit against a table. **Observe** the sound. Try hitting the spoon against the table without holding your ears closed. Describe the different sounds. How do you think vibrations of the spoon affected the sound you heard?

Science in a Snap! **Observe Static Electricity**

Place an empty aluminum can on its side on a flat surface. Rub a piece of wool cloth on a balloon. Hold the balloon close to the can without touching it. **Observe** what happens to the can. Next, rub the wool cloth on a second balloon and put that balloon on the table. Rub the first balloon with wool again. **Predict** what will happen if you bring the two balloons near each other. Try it! Was your prediction supported? What force acted on the balloon and the can?

Investigate Mixtures

**Question** How can you separate the materials in a mixture using their physical properties?

Science Process Vocabulary

observe verb

When you **observe,** you use your senses to tell about the properties of an object.

infer verb

When you **infer,** you use what you know and what you observe to draw a conclusion.

When I mix sugar and water, the sugar seems to disappear. I can infer that it dissolves.

Materials

safety goggles

mixture

hand lens

3 plastic cups

magnet

plastic bag

plastic mesh

filter paper

What to Do

1 Put on your safety goggles. Use the hand lens to **observe** the mixture of iron filings, pebbles, sand, salt and water. Record your observations in your science notebook.

2 Place the magnet in a plastic bag. Then put the bag with the magnet into the mixture. Stir the mixture with the magnet to attract the iron filings. Keep stirring until there are no more filings in the mixture. Remove the plastic bag and magnet and place them in a cup.

3 Put the plastic mesh on top of another cup. Pour the remaining mixture through the mesh into the cup. Set aside the mesh and the pebbles that collected on it.

What to Do, continued

4 Put the filter paper over an empty cup as shown. Pour the mixture you collected in step 3 into the filter. Wait until all the liquids goes through the filter. Then set aside the filter paper and the sand that collected on it.

5 Observe the liquid you collected in the cup. Record your observations. How many different materials do you think are in the liquid? Write your **prediction** in your science notebook.

6 Place the liquid in sunlight until all the liquid evaporates. Observe the empty cup. Record your observations.

Record

Write in your science notebook. Use a table like this one. Record your observations.

Mixture

What I Observed	Observations
Mixture at start	
Liquid in step 5	
Cup after liquid evaporates	

Explain and Conclude

1. What materials made up the original mixture?

2. In step 6, what did you **observe** in the cup after all the liquid evaporated? **Infer** where the material came from. Does this result support your **prediction?**

3. How did you separate each of the materials from the mixture? What physical property of each material allowed you to separate it from the mixture?

What components of this mixture can you identify?

Math
in Science

Measuring Volume Accurately

When doing investigations, scientists often make measurements. They might measure mass, volume, length, temperature, or other values. Measurements can help scientists compare objects, find patterns, or find out if something has changed.

Scientists must measure accurately. If they make an error while measuring, their data will be wrong. Making sure that the data are recorded correctly is also important. Incorrect data can affect how the results of an investigation are interpreted. If the data are incorrect, the conclusions probably are incorrect too.

The first step scientists take to make accurate measurements is to decide what data they want to collect. Next they decide how precise their measurements must be. Then they choose the tool that is best suited for the task.

Measuring the Volume of a Liquid

Suppose you want to find the volume of a liquid. You can measure with a graduated cylinder or a measuring cup. A graduated cylinder will give you a more precise measurement. If you do not need an exact amount of a liquid, you may choose to use a measuring cup.

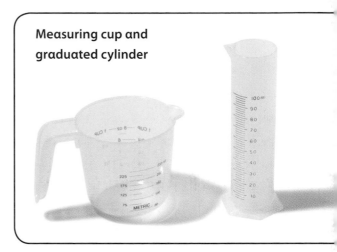

Measuring cup and graduated cylinder

To find the volume of a liquid with a graduated cylinder, follow these steps:

1. Place the graduated cylinder on a flat surface.

2. Move so that your eyes are even with the top of the liquid in the cylinder.

3. In a graduated cylinder, the liquid may curve. Look at the bottom of the curve in the middle of the graduated cylinder.

4. Match that point with a number on the scale on the graduated cylinder. Most graduated cylinders measure liters (L) or milliliters (mL). When you measure, make sure you know what each mark shows.

Measuring the Volume of a Solid

Using a Graduated Cylinder You can also use a graduated cylinder to measure the volume of a solid object.

1 Fill the graduated cylinder with enough water to cover the object. Measure the volume of the water.

2 Place the solid in the graduated cylinder. The water level will rise.

3 Measure the volume of the water and the solid.

4 Subtract the volume of the water from the volume of the water and the solid.

	Volume		
	Water	Water and Rock	Rock
Rock	30 mL	55 mL	25 mL

Using a Metric Ruler If the solid you are measuring is a square or a rectangular block, measure the length, width, and height. Then multiply the values to find the volume in cubic centimeters, or cm^3.

	Length	Width	Height	L x W x H	Volume
Block	15 cm	9 cm	4 cm	15 x 9 x 4	540 cm³

SUMMARIZE

What Did You Find Out?

1 Why is it important to make and record accurate measurements?

2 What are two ways to measure the volume of a solid?

Practice Measuring Volume

Measure the volume of a solid in 2 ways.

1 Make a small clay cube. It should be small enough to fit in a graduated cylinder.

2 First find the volume of the cube by using the formula: L x W x H.

3 Then find the volume using a graduated cylinder.

4 **Compare** the 2 volume measurements you found.

Investigate Properties of Water

Question How does changing the state of water affect its physical properties of mass and volume?

Science Process Vocabulary

measure verb

When you **measure**, you use tools to find out how much or how many.

compare verb

When you **compare**, you tell how objects or events are alike and different.

These icicles are different shapes and sizes.

Materials

graduated cylinder with ice

balance

gram masses

Calculating Mass Worksheet

plastic cup

168

What to Do

1 Remove the plastic wrap from the graduated cylinder. **Measure** the volume of the ice in the graduated cylinder. Record the volume in your science notebook. Put the plastic wrap back on the graduated cylinder to prevent evaporation.

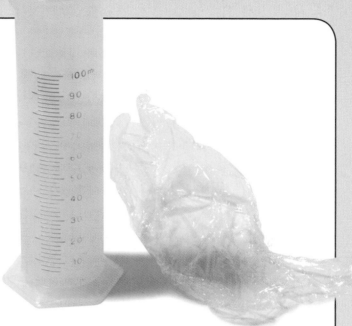

2 Use the balance and gram masses to measure the mass of the graduated cylinder, ice, and plastic wrap. Record the mass on the Calculating Mass Worksheet.

What to Do, continued

3 Let the ice melt. Then repeat steps 1–2 with the graduated cylinder, water, and plastic wrap.

4 Pour all the water out of the graduated cylinder into the plastic cup. Use the balance and gram masses to measure the mass of the empty graduated cylinder and the plastic wrap. Record the mass on the Calculating Mass Worksheet.

5 Use the Worksheet to calculate the mass of the ice and of the water. Record your **data.** Record your calculations in your science notebook.

Record

Write in your science notebook.
Use a table like this one.

Physical Properties of Water

What I Observed	Before Melting	After Melting
Volume (mL)		
Mass (g)		

Explain and Conclude

1. Compare your **measurements** of the water before and after melting. Did the water's volume and mass change after it melted?

2. What can you **conclude** about how changing the state of water affects the physical properties of mass and volume?

Think of Another Question

What else would you like to find out about how the change in state of water affects its physical properties?

Investigate Properties of Matter

Question

How do the properties of matter change as it changes state?

Science Process Vocabulary

Materials

predict verb

When you **predict,** you tell what you think will happen.

> I predict that the icicle will melt in the sunlight.

ice cube

plastic cup

infer verb

When you **infer,** you use what you know and what you observe to draw a conclusion.

> I infer that it rained recently.

plastic wrap

What to Do

1 Put an ice cube in the plastic cup. **Observe** as many of the properties of the ice cube as possible. Record your observations in your science notebook.

2 Place the cup with the ice cube in a sunny spot. **Predict** how the properties of the ice will change after 1 hour. Record your predictions.

3 Observe the cup and its contents after 1 hour. Record your observations.

What to Do, continued

4 Stretch plastic wrap across the top of the cup. Place the cup in a sunny spot. Predict how the contents of the cup will change after 2 hours. Record your predictions.

5 After 2 hours, observe the properties of the contents of the cup. Record your observations.

6 Remove the plastic wrap from the cup. Predict what will happen to the contents of the cup. Allow the uncovered cup to sit in a sunny location for 3 days. Then observe the contents of the cup again. Record your observations.

Record

Write in your science notebook.
Use a table like this one.

Contents of the Cup

	Prediction	Observations
Start		
1 hour in sunlight		
2 hours in sunlight (covered)		

Explain and Conclude

1. What happened to the properties of the ice as it sat in the sunlight? How did the properties of the water change as the cup sat in the sunlight for 3 days?

2. **Infer** what caused these changes.

3. What states of matter did you **observe** in this activity? Is changing state a physical or chemical change?

Think of Another Question

What else would you like to find out about the changing states of matter? How could you find an answer to this new question?

Niagara Falls, New York

Investigate Rusting

Question What happens when steel wool in a test tube rusts?

Science Process Vocabulary

variable noun

A **variable** is something that could change in an investigation. You only change one variable in an experiment.

> I will change the amount of steel wool I use in each test tube.

Materials

spray bottle

steel wool

straw

3 test tubes

3 cups of water

tape

Do an Experiment

Write your plan in your science notebook.

Make a Hypothesis

In this investigation, you will place 2 test tubes with different amounts of steel wool upside down in cups of water. A chemical reaction will take place in which the steel wool reacts with the oxygen in the air to make rust. What do you think will happen to the volume of water in the cups and test tubes as the steel wool rusts? Write your **hypothesis.**

Identify, Manipulate, and Control Variables

Which variable will you change?
Which variable will you observe or measure?
Which variables will you keep the same?

What to Do

1 Use the spray bottle to dampen a small piece of steel wool. Push the steel wool to the bottom of a test tube with a straw.

2 Place the test tube upside down in a cup of water. Tape the test tube to the side of the cup. Label the cup **1.**

What to Do, continued

3 Change your **variable.** Select an amount of steel wool that is different from the amount in the test tube in cup 1. Dampen the steel wool with the spray bottle and push it to the bottom of the test tube. Assemble the test tube and a cup as in step 1. Label the cup **2.**

4 Place an empty test tube upside down in another cup of water. Tape the tube to the cup and label the cup **3.** This empty test tube is your control.

5 Set the 3 cups with the test tubes in a place where they can stay undisturbed for 3 days. Every day, **observe** the steel wool and the amount of water in the test tubes. Record your observations.

Record

Write in your science notebook.
Use a table like this one.

SCIENCE
notebook

Observations of Steel Wool and Water Level

	Start	1 Day	2 Days	3 Days
Cup 1				
Cup 2				
Cup 3				

Explain and Conclude

1. What changes did you **observe** in the steel wool?
   What caused these changes?

2. How did the water levels in the cups and test tubes change?
   Use this evidence to **infer** what happened to the oxygen
   in the test tubes that caused this change.

3. **Compare** your results with other
   groups. Explain why your results might
   have been different.

Think of Another Question

What else would you like to find out about
rusting? How could you find
an answer to this new question?

Over time the metal in
this bike has rusted.

Science and Technology

Designing and Building Better Products

Scientists and engineers work together to improve the products we use each day. Scientists develop new materials that are stronger and more durable. Engineers test the designs of machines and electrical systems. Scientists also design products that help people take good care of their health, such as the toothbrush in the picture below.

A toothbrush is a simple product, but many people over hundreds of years were involved in designing it. Long ago, people used twigs, bird feathers, or animal bones to clean their teeth.

▶ This is a model of a siwak toothbrush, one of the first kinds of toothbrushes. It was made from tree bark.

▼ The electric toothbrush was designed to help people take better care of their teeth and gums.

In the 1400s, people made toothbrushes by putting holes in sticks made of wood or animal bone. They pulled animal hair through the holes to make bristles. This design was not ideal because the bristles often fell out or contained bacteria that could make people sick. Animal bones were still used for the handles of many toothbrushes in the 1800s.

▲ This toothbrush has a bone-handle and boar-bristle brush. It is from 1869.

▼ The handle of this toothbrush, from 1870–1920, looks like animal ivory.
But the material it is made from was developed by scientists.

By the early 1900s, inventors were using plastics to make products that were more lightweight and durable. They started to make the bristles and handles of toothbrushes out of plastic. The inventors combined research and the new technology to design a toothbrush that would help people take better care of their teeth and gums.

This brush has a combination bone and plastic
handle. It was made in the late 1800s or early 1900s.

Identifying a Design Problem Teams work together to test existing products to see how well they work. The team finds out what changes people would like in the product. The team uses the information to identify the design problem they need to work on in order to make the product better or to make a new product.

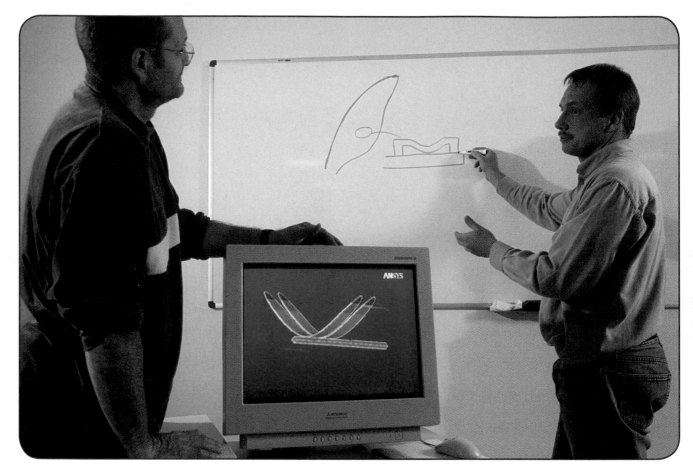

These scientists are discussing the design of a dragonfly-shaped micro air vehicle, or microdrone.

Investigating Solutions Next, the team suggests possible solutions. They talk about ways the product might be made differently, fixed, or improved. They might research the different designs that have been used in the past. This way, they can learn from the experiences of other scientists and not repeat their mistakes. They might also discuss different materials or new technology that would make the product better. The team considers whether they need to make a new product or improve an existing one.

Making a Plan An important step in solving a design problem is to develop a plan. First, the team must identify the purpose of their product. What problem will it solve or what need will it meet? What will it look like? The team decides which materials they will use and which steps they will take. They think about possible problems that might affect their design. They plan ahead. They make sure their plan is safe to test.

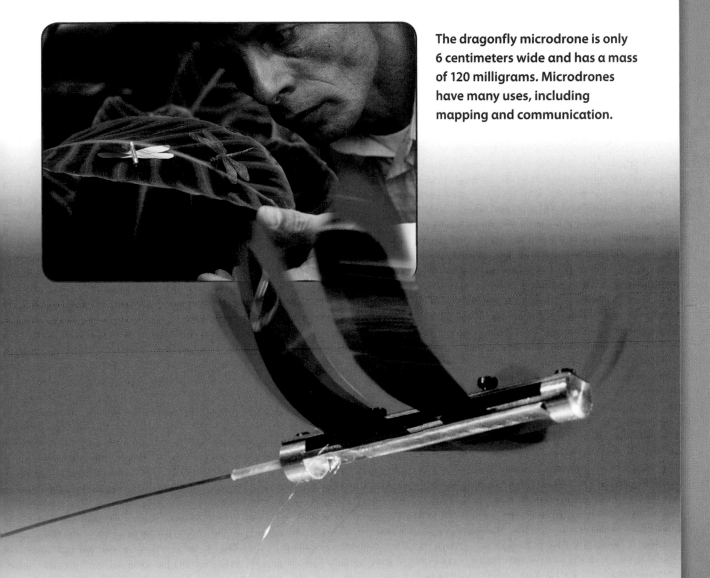

The dragonfly microdrone is only 6 centimeters wide and has a mass of 120 milligrams. Microdrones have many uses, including mapping and communication.

Testing the Design To make sure that the new design works, the team builds an example of the product called a prototype. They test the prototype to find out whether it will work in different conditions and is safe to use. The team does multiple tests to be sure their results are accurate.

Analyze the Results The team analyzes the results of their tests. First, they make sure there are no errors in the data. Then they analyze how well the prototype worked and whether it solved the problem they identified. Finally, scientists will report the results of their tests and share the new design of the product.

Researchers use a microscope to observe the wing of the dragonfly microdrone.

SUMMARIZE
What Did You Find Out?

1. Why do scientists and others that work on new products research designs that have been used in the past?

2. What are four things that a team might consider when they make a plan for a new product?

3. Why do scientists build prototypes of new designs?

Design a Better Grocery Bag

A grocery store chain wants to help the environment. They decide to develop a new grocery bag that people can reuse many times. The bag must be large enough to hold large grocery items, and it must be strong enough to hold heavy items. Help the store chain by designing a new grocery bag. Follow these steps:

1 Identify any problems with the existing grocery bag. State the problem you would like to solve.

2 Investigate solutions. What will you make differently, fix, or improve? Research existing grocery bags to help you think about your new design. Draw your design.

3 Make a plan. Identify the purpose of the product. What problem will it solve? Think about the materials you will need and the steps you will take. Make sure your plan is safe to test.

4 Use the available materials to build a prototype of your grocery bag. Test the bag to see how well the design works. Record your results.

5 Analyze the results of your tests. Check for possible errors. Decide whether you have any further changes to make.

6 Share your final design with others. Explain how your design will be helpful to people who use it.

Investigate Magnets and Forces

Question How can you use a magnet to move a car without touching it?

Science Process Vocabulary

Materials

observe verb

When you **observe**, you use your senses to learn about an object or event.

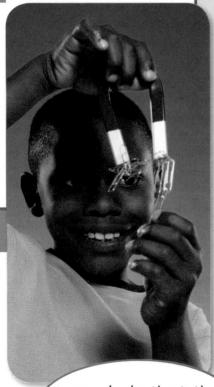

conclude verb

You **conclude** when you use data from an investigation to come up with a decision or answer.

I conclude that the force of the magnet attracts the objects.

2 bar magnets

toy car

tape

ruler

index card

What to Do

1 Tape 1 bar magnet to the top of the toy car. The south pole of the magnet should point toward the front of the car.

2 Align the back of the car with the 0 end of the ruler. Hold the second magnet above the ruler with the S end of the magnet at the 5 cm mark. Slowly move second magnet toward the back of the car. **Observe** what happens to the car and how close the magnet was to it when it moved. Record your observations in your science notebook. Then do 2 more trials.

3 Repeat step 2 with the north pole of the free magnet pointing toward the back of the car.

What to Do, continued

4 Fold an index card in half. Tape one end of the index card to the table, as shown in the photo.

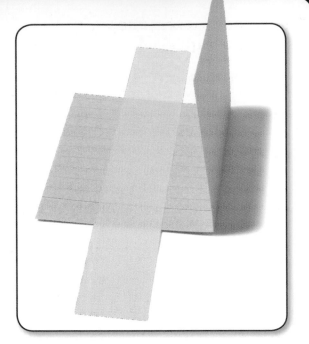

5 Place the toy car about 10 cm away from the card. Point the north pole of the magnet toward the car. Use the magnet to push the car toward the card. Observe what happens to the car when it hits the card. Record your observations. Then do 2 more trials.

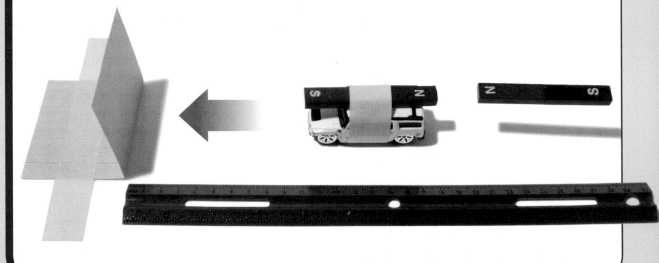

Record

Write in your science notebook.
Use tables like these.

Toy Car and Bar Magnet

Trial	Which Poles Were Facing Each Other?		How Close Magnet Was to Car (cm)	Observations
	Magnet on Car	Free Magnet		
1	North	South		
2				

Toy Car and Index Card

Trial	Observations
1	
2	

Explain and Conclude

1. What happened when you moved the magnet toward the car in steps 2 and 3? What force caused this to happen?

2. What happened to the motion of the car when it hit the index card? What force caused this to happen?

Think of Another Question

What else would you like to find out about using a magnet to move objects without touching them? How could you find an answer to this new question?

Investigate Forces

Question What happens to the motion of a toy car when it rolls onto different surfaces?

Science Process Vocabulary

hypothesis noun

You make a **hypothesis** when you state a possible answer to a question that can be tested by an experiment.

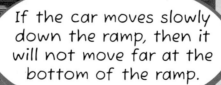

If the car moves slowly down the ramp, then it will not move far at the bottom of the ramp.

measure verb

When you **measure,** you find out how much or how many.

Materials

safety goggles

spring scale

toy car

2 books

cardboard

tape

ruler

meterstick

fine sandpaper

rough sandpaper

burlap

Do an Experiment

Write your plan in your science notebook.

Make a Hypothesis

In this investigation, you will observe a toy car rolling down a cardboard ramp onto different materials: fine sandpaper, rough sandpaper, and burlap. How will each material affect the motion of the toy car? Write your **hypothesis.**

Identify, Manipulate, and Control Variables

Which variable will you change?
Which variable will you observe or measure?
Which variable will you keep the same?

What to Do

1   Put on your safety goggles. Hold the spring scale in front of you. Attach the toy car to the spring scale. **Measure** the pull of gravity on the toy car. Record the **data** in your science notebook.

2   Stack 2 books on the floor. Tape one end of the cardboard to the books and the other end to the floor to make a ramp.

What to Do, continued

3 Place the toy car at the top of the ramp. Let go of the car so that it rolls down the center of the ramp. Do not push the car. When the car comes to a complete stop, use a ruler or meterstick to measure how far the car rolled past the end of the ramp. Record the distance.

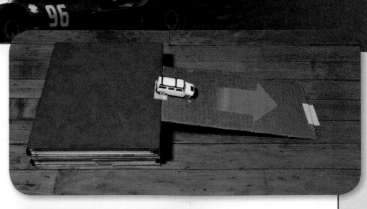

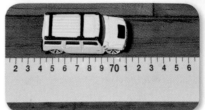

4 Repeat step 3 two more times.

5 Choose fine sandpaper, rough sandpaper, or burlap and place it at the end of the ramp. Tape the corners of the material to the floor so that it lays flat.

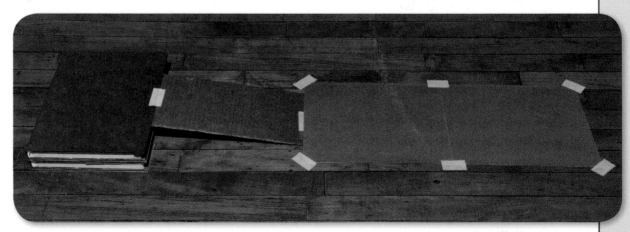

6 Place the car at the top of the ramp and let go. Measure and record how far the car rolls past the end of the ramp. Then repeat this step two more times.

7 Repeat steps 5 and 6 with a different material.

Record

Write in your science notebook.
Use a table like this one.

Distance Traveled By Toy Car

Material	Trial 1	Trial 2	Trial 3
Bare floor			

Explain and Conclude

1. What force caused the car to move down the ramp?

2. **Compare** your results with other groups. Which material caused the greatest change to the distance travelled by the toy car?

3. What can you **conclude** about which surface exerted the greatest force on the car? How do you know?

Think of Another Question

What else would you like to find out about how different surfaces affect the motion of objects? How could you find an answer to this new question?

Investigate Magnets

Question How will magnets attract and repel other magnets and materials?

Science Process Vocabulary

predict verb

You **predict** when you say what you think will happen.

> I predict that the magnet will attract the iron filings.

operational definition noun

When you make an **operational definition,** you use your own words to tell what something can do.

> A paper clip is a metal wire that is bent so that it can hold papers together.

Materials

2 large bar magnets

ruler

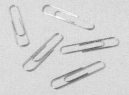
1 small bar magnet

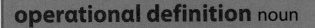
5 paper clips

rubber band

What to Do

1 Place 2 bar magnets about 10 cm apart with their north poles facing each other. **Predict** what will happen if you move the magnets toward each other. Record your prediction in your science notebook. Slowly push the magnets toward each other. **Observe** and record what happens.

2 Repeat step 1 with the south poles of the magnets facing each other, and then again with a north pole facing a south pole.

3 Predict what will happen if you hold the north pole of a magnet near the paper clips. What will happen if you hold the south pole near the paper clips? Record and test your predictions.

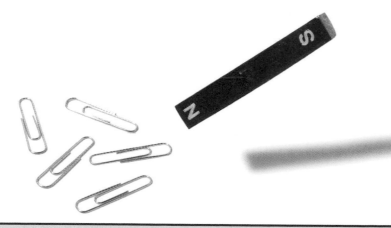

What to Do, continued

4 Place a paper clip just past the end of the ruler, as shown. Place a bar magnet alongside the ruler so that its north pole is 10 cm from the paper clip. Predict how close the magnet will get before the magnet's force affects the paper clip. Record your predictions.

5 Slowly move the magnet toward the paper clip. Record your observations.

6 Hook a rubber band to a paper clip. Hold the south pole of the small bar magnet near the paper clip so that it attracts the paper clip. Predict what will happen if you gently pull the paper clip away from the magnet. Then very slowly pull the rubber band. Record your observations.

7 Repeat steps 4–6 with a different magnet.

Record

Write or draw in your science notebook.
Use a table like this one.

Set-Up	Magnets			
	Prediction		**Observation**	
	Magnet 1	Magnet 2	Magnet 1	Magnet 2
North poles facing				
South poles facing				

Explain and Conclude

1. Did your results support your **predictions** about the way magnets would attract and repel other magnets? Explain.

2. What effect did using different poles of the magnet have on the paper clips?

3. Use your results to write an **operational definition** for a magnet.

Think of Another Question

What else would you like to find out about magnets? How could you find an answer to this new question?

Magnets are used in many tools and machines, such as this MRI machine.

Investigate Magnetic Fields

Question How will iron filings line up when placed near different magnets?

Science Process Vocabulary

compare verb

When you **compare**, you tell how objects or events are alike and different.

conclude verb

You **conclude** when you use information, or data, from an investigation to come up with a decision or answer.

I can conclude that the paper clip is magnetic but the craft stick is not.

Materials

safety goggles

iron filings

tape

2 bar magnets

2 donut magnets

horseshoe magnet

What to Do

1 When a magnet is placed near iron filings, the iron filings will line up to show the magnetic field of the magnet.

2 Put on your safety goggles. Tape down the plastic bag with iron filings.

3 Hold a bar magnet 1 cm above the plastic bag, near the middle of the iron filings. **Observe** how the iron filings line up. Record your observations in your science notebook.

What to Do, continued

4 **Predict** how the iron filings will line up when you hold another bar magnet 1 cm above the plastic bag so that the north poles of the magnets are facing each other. The poles of the magnets should be about 1 cm apart. Record and test your prediction.

5 Predict how the iron filings will line up if the north and south poles of the bar magnets face each other. Record and test your predictions. Observe the iron filings and draw your observations.

6 Choose 2 more magnets to test together. **Predict** how the iron filings will line up if you hold the magnets above the plastic bag. Record and test your prediction. Record your observations.

Record

Write and draw in your science notebook.
Use a table like this one.

Iron Filings

Magnets	Prediction	Observations
1 bar magnet		
2 bar magnets N – N		

Explain and Conclude

1. Did your results match your **predictions?** Explain.

2. **Compare** the pattern of iron filings in steps 4 and 5. Explain any differences.

3. Use your **observations** to **conclude** whether each pair of magnets attracted or repelled each other.

Think of Another Question

What else would you like to find out about magnetic fields? How could you find an answer to this new question?

Arctic Circle, Alaska

The Aurora Borealis occurs because particles from the sun react with Earth's magnetic force.

Investigate Energy

Question

Which forms of energy can you observe by holding a foil spiral over a lamp?

Science Process Vocabulary

analyze verb

When you **analyze** data, you examine them closely to find out what they mean.

infer verb

You **infer** when you use what you know and what you observe to draw a conclusion.

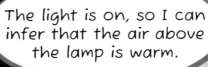

The light is on, so I can infer that the air above the lamp is warm.

Materials

safety goggles

lamp

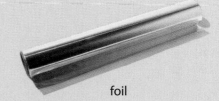
foil

ruler

scissors   thread

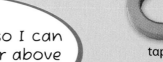
tape

What to Do

1 Put on your safety goggles. Place a small lamp on the table. Turn on the lamp. Be sure not to look directly at the light or touch the bulb. **Observe** the forms of energy that the lamp uses and produces. Record your observations in your science notebook.

2 **Measure** a circle with a diameter of 12 cm on a piece of foil. Cut out the circle. Then cut the circle into a spiral, as shown.

What to Do, continued

3 Measure and cut a piece of thread that is 20 cm long. Use a small piece of tape to attach one end of the thread to the center of the foil spiral.

4 Hold the thread above the lamp so that the foil spiral is about 5 cm above the light bulb. Make sure the foil does not touch the bulb. Hold the thread very still and **observe** the foil spiral. Record your observations.

5 **Analyze** your observations of the foil spiral. What form of energy causes a change in the foil? What form of energy does the spiral have?

Record

Write in your science notebook.
Use a table like this one.

	Observations	Forms of Energy It Uses	Forms of Energy It Has or Produces
Lamp			
Foil spiral			

Lamp and Foil Spiral

Explain and Conclude

1. What forms of energy does the lamp use? What forms of energy does the lamp produce?

2. What happened to the foil spiral when you held it over the lamp? **Infer** what caused this to happen.

Think of Another Question

What else would you like to find out about forms of energy? How could you find an answer to this new question?

Pacific Grove, California

What types of energy change occur when the waves push against the rocks?

Investigate Heat Conductors

Question Will a plastic ruler, a wooden ruler, or a metal ruler conduct heat best?

Science Process Vocabulary

measure verb

When you **measure,** you use tools to find out how much or how many.

compare verb

When you **compare,** you tell how things are alike and different.

Materials

3 strip thermometers

metal ruler

wooden ruler

plastic ruler

3 pitchers of water

graduated cylinder

jar

tape

stopwatch

What to Do

1 Place 1 strip thermometer on your desk. **Observe** its color. Then press your finger on the top of the strip thermometer for about 5 seconds. Observe the part of the strip that you touched. Record your observations in your science notebook.

2 **Infer** what caused the color change on the strip thermometer. Write in your science notebook.

3 Remove the protective backing from each strip thermometer. Place a strip about 7.5 cm from the end of the metal, wooden, and plastic rulers. Press each strip so that it sticks to the ruler.

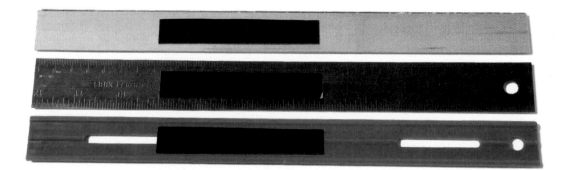

4 Fill the jar with 250 mL of warm water.

5 Tape each ruler to the side of the jar. Observe the strip thermometers and record your observations.

6 Observe the strip thermometers every 30 seconds for 3 min. Use the stopwatch to **measure** your time.

7 Remove the rulers from the jar. Choose water of a different temperature. Fill the jar with 250 mL of the water then repeat steps 5 and 6.

Record

Write or draw in your science notebook.
Use a table like this one.

Observations with Warm Water

Time	Wooden Ruler	Plastic Ruler	Metal Ruler
Start			
30 seconds			
1 minute			

Explain and Conclude

1. On which ruler did the color of the strip thermometer change the most? Was there any difference when you used water of a different temperature?

2. Based on your observations, **infer** what caused the color of the strips to change. Explain any differences you may have observed in the strips.

3. **Compare** your results with other groups. What are some reasons your results might be different?

Think of Another Question

What else would you like to find out about materials that conduct heat? How could you find an answer to this new question?

Investigate Vibrations

Question How does the tightness of a string affect the sound it makes when you pluck it?

Science Process Vocabulary

observe verb

When you **observe,** you use your senses to learn about an object or event. You can use your sense of hearing to observe sound.

infer verb

When you **infer,** you use what you know and what you observe to draw a conclusion.

I can infer that thicker strings make different sounds than thinner strings.

Materials

safety goggles

ruler

fishing line

scissors

2 paper clips

cup with hole

What to Do

1 Put on your safety goggles. Cut a piece of fishing line that is 30 cm long. Tie a paper clip to one end of the fishing line.

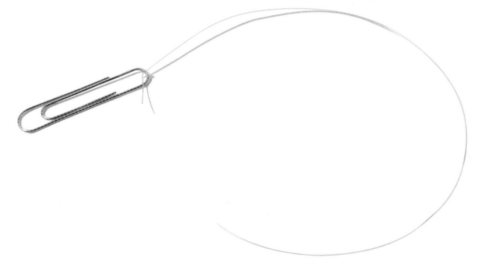

2 Push the fishing line through the hole in the cup so that the paper clip is on the outside of the cup, as shown. Then tie another paper clip to the loose end of the fishing line.

What to Do, continued

3 Work with a partner. Have your partner hold the cup while you hold the paper clip at the end of the fishing line. Gently pull the fishing line so that it tightens. Pluck the fishing line. **Observe** the vibration of the fishing line and the sound it makes. Record your observations in your science notebook.

4 **Predict** how the sound will change if you hold the fishing line tighter. Record your prediction. Then pull the line tighter and pluck it. Observe its vibration and the sound it makes. Record your observations.

5 Repeat step 4, but this time hold the line looser. Did you observe a change in the sound?

Record

Write or draw in your science notebook.
Use a table like this one.

Sounds Made by Plucking Fishing Line

Tightness of Fishing Line	Prediction	Observation
Start		
Tighter		
Looser		

Explain and Conclude

1. Did your observations support your **predictions?** Explain.

2. What change did you **observe** in the vibrations of the fishing line when you pulled it tighter? How did the sound change?

3. **Infer** how tightening the fishing line more would affect the sound it makes.

Think of Another Question

What else would you like to find out about the tightness of a string and pitch? How could you find an answer to this new question?

Investigate Pitch

Question How can you change the pitch of the sound made by a straw instrument?

Science Process Vocabulary

Materials

plan noun

When you make a **plan** to answer a question, you list the materials and steps you need to take.

> I will first cut a point at the end of the straw. Then I will blow through the straw.

share verb

When you **share** results, you tell or show what you have learned.

4 straws

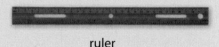

ruler

scissors

hole punch

clear tape

What to Do

1 Use a metric ruler to **measure** and mark a line 1 cm from the end of the straw. Cut the straw to make a 1 cm point.

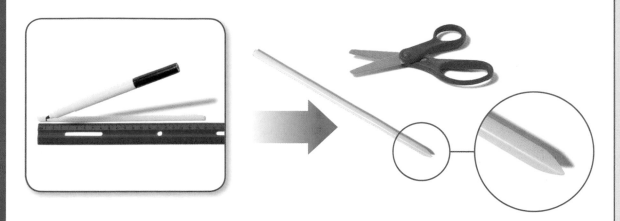

2 Push the pointed end of the straw between your thumb and forefinger to flatten it. You have made a musical instrument!

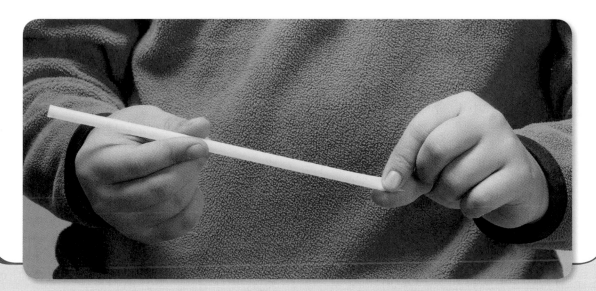

What to Do, continued

3 Purse your lips around the straw where the straw was cut. Lightly push down on the straw with your lips as you blow through it. Blow until you produce a sound. You may have to try several times until you figure out the best way to hold your lips.

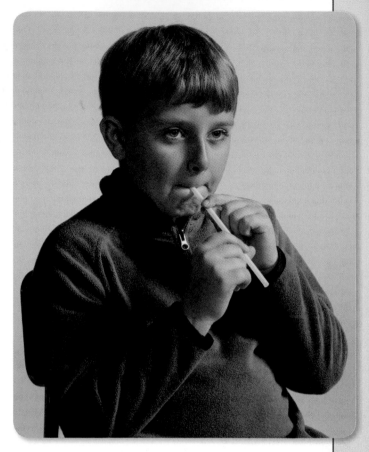

4 **Observe** the sound that you make with the straw. Pay particular attention to the pitch of the sound. Also observe how the pointed part of the straw vibrates.

5 Think about ways you could change your instrument to produce sounds with different pitches. You can use any of the materials in the Materials list. Make a **plan.** Try a few of your ideas. Record the changes you make and how each change affects the pitch of the sound.

Record

Write in your science notebook.
Use a table like this one.

Changes in Straw and Pitch

How I Changed the Straw	What Happened to the Pitch of the Sound

Explain and Conclude

1. Use your **observations** in step 4 to **infer** what caused your straw instrument to make a sound.

2. How did you change your instrument to make different sounds?

3. **Share** your results with others. Look for patterns. Which kinds of changes make sounds with a higher pitch? Which make sounds with a lower pitch?

Think of Another Question

What else would you like to find out about changing the pitch of a sound? How could you find an answer to this new question?

Investigate Circuits

 **How are series circuits and parallel circuits different?**

Science Process Vocabulary

predict verb

You **predict** when you tell what you think will happen.

> I predict that the lamp will produce light if I plug it in and turn the switch.

compare verb

When you **compare,** you tell how objects or events are alike and different.

> The flashlight is brighter than the lamp.

Materials

safety goggles

battery

battery holder

2 light bulbs and holders

4 wires

What to Do

1 Put on your safety goggles. Use 3 wires to make a series circuit. Connect the wires to the light bulb holders and the battery holder. Then put the battery in the battery holder.

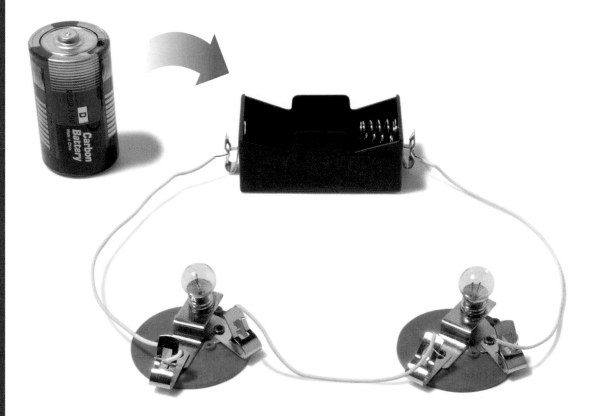

2 **Observe** what happens when all the parts of the circuit are connected. Record your observations in your science notebook. Draw a diagram that shows the parts of your series circuit.

3 What do you think will happen if you unscrew 1 light bulb? Record your **prediction.** Then remove a light bulb from the series circuit and record what happens.

What to Do, continued

4 Take apart the series circuit. Use 4 wires to make a parallel circuit. Record what happens when all the parts of the circuit are connected. Draw a diagram of your parallel circuit.

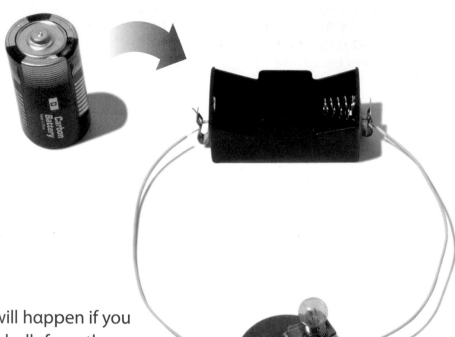

5 Predict what will happen if you remove 1 light bulb from the parallel circuit. Record your prediction. Then remove a light bulb from the parallel circuit and record what happens.

Record

Write and draw in your science notebook.
Use a table like this one.

Series and Parallel Circuits

Type of Circuit	Observations: Circuit With All Parts Connected	Prediction: What Will Happen If a Light Bulb Is Removed?	Observations: Circuit With Light Bulb Removed
Series			
Parallel			

Series Circuit Diagram

Parallel Circuit Diagram

Explain and Conclude

1. Did your **observations** support your **predictions?** Explain.

2. **Compare** your observations of the two circuits after you removed a light bulb. How is the parallel circuit different from the series circuit?

3. Why do you think parallel circuits are more commonly used than series circuits?

Think of Another Question

What else would you like to find out about series and parallel circuits? How could you find an answer to this question?

Complex circuits like this one can carry electrical energy to hundreds or even thousands of homes.

Investigate Conductors and Insulators

Question Which materials allow electricity to flow?

Science Process Vocabulary

classify verb

When you **classify** objects, you put them in groups according to their characteristics.

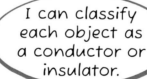

> I can classify each object as a conductor or insulator.

operational definition noun

When you make an **operational definition,** you use your own words to tell what something can do.

> An electrical insulator is a material that does not allow electricity to flow through it.

Materials

safety goggles   battery   battery connector

3 wires   electrical tape

scissors   buzzer

paper clip   lemon juice   chopstick

washer   water   pencil

plastic spoon   crayon   aluminum foil

What to Do

1 Put on your safety goggles. Make an electrical circuit. Use 3 wires to attach the battery connector and buzzer, as shown. Wrap a small piece of electrical tape around the exposed connections. Make sure the long, stripped parts of the wire are not connected to anything. Attach the battery to the battery connector.

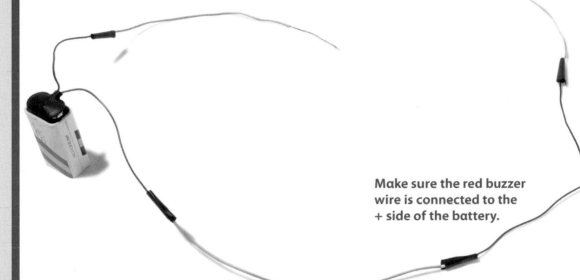

Make sure the red buzzer wire is connected to the + side of the battery.

2 Test your circuit. Holding the insulated part of the wire only, touch the end of each wire to the paper clip. The paper clip is a conductor so electricity should flow through the circuit and the buzzer should make a sound. Record your **observations** in your science notebook.

What to Do, continued

3   Select a different material to complete your circuit. **Predict** whether the material is a conductor or an insulator. Record your prediction.

4   If you chose to test a solid object, touch the wires to the object. If you chose to test a liquid, carefully place the wires in the cup of liquid. Record your observations.

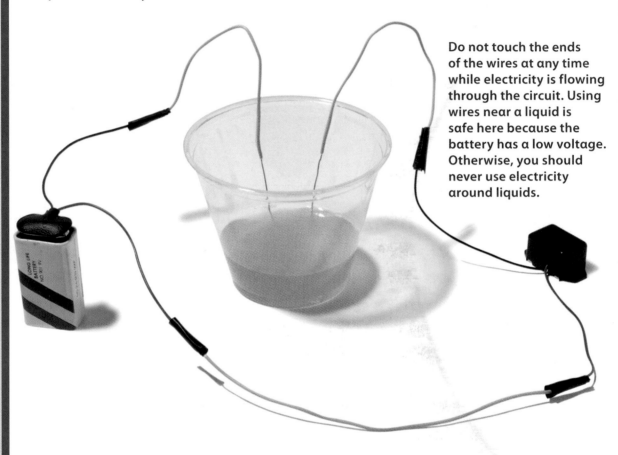

Do not touch the ends of the wires at any time while electricity is flowing through the circuit. Using wires near a liquid is safe here because the battery has a low voltage. Otherwise, you should never use electricity around liquids.

5   Repeat steps 3–4 with 3 different materials. Record your predictions and observations.

6   **Classify** each of the materials you tested as a conductor or an insulator. Write in your science notebook.

Record

Write in your science notebook.
Use a table like this one.

Classification of Materials

Material	Prediction: Conductor or Insulator?	Observation: Did the Buzzer Make a Sound?	Classification: Conductor or Insulator?
Paper clip		yes	conductor

Explain and Conclude

1. Which materials were conductors? How could you tell?

2. Write an **operational definition** for an electrical conductor.

3. **Compare** your results with other groups. How are the materials you classified as conductors alike?

Think of Another Question

What else would you like to find out about electrical conductors and insulators? How could you find an answer to this new question?

Wires are used to conduct electricity to homes and other buildings.

225

Do Your Own Investigation

Question **Choose one of these questions, or make up one of your own to do your investigation.**

- What happens if you separate the colors in black ink from different brands of markers?
- If you mix baking soda and vinegar, how will the temperature of the vinegar affect the rate of the chemical change?
- Which will roll down a ramp faster, a full water bottle or an empty water bottle?
- How can you change the magnetic strength of a magnetized nail?
- What happens when you add hot water to a container of cold water?
- How do the sounds compare when you hit wooden blocks and plastic blocks?
- How does the number of light bulbs in a parallel circuit affect the brightness of each bulb?

Science Process Vocabulary

I will conduct an experiment to test my hypothesis that all black inks contain the same colors.

hypothesis noun

You make a **hypothesis** when you state a possible answer to a question that can be tested by an experiment.

Open Inquiry Checklist

Here is a checklist you can use when you investigate.

- ☐ Choose a **question** or make up one of your own.

- ☐ Gather the materials you will use.

- ☐ If needed, make a **hypothesis** or a **prediction.**

- ☐ If needed, identify, manipulate, and control **variables.**

- ☐ Make a **plan** for your **investigation.**

- ☐ Carry out your **plan.**

- ☐ Collect and record **data. Analyze** your data.

- ☐ Explain and **share** your results.

- ☐ Tell what you **conclude.**

- ☐ Think of another question.

Scientists can test the colors in ink to find out what type of pen was used to write a note or letter.

Write Like a Scientist

Write About an Investigation

The Colors in Black Ink

The following pages show how one student, Cara, wrote about an investigation. As she read about the properties of matter, Cara became interested in why objects have different colors. She decided to do an investigation to find out what other colors are in black ink. Here is what she thought about to get started:

- Cara read that there are many colors present in black ink. She decided to test two different brands of black markers to see whether their inks contain the same colors.

- She needed to use materials that were safe and easy to obtain. She would design her investigation based on the materials she could use.

- Cara decided to draw on coffee filters with different black markers. She would dip the coffee filters in water to separate the colors in the ink.

- She would compare the colors that were visible in the ink of each marker.

Model

Question

What happens if you separate the colors in black ink from different brands of markers?

> Choose a question that can be answered using materials that are safe and easy to obtain.

Materials

2 plastic cups

water

graduated cylinder

scissors

white coffee filter

2 brands of black, water-soluble markers

tape

2 craft sticks

> List the number of each item you will need if you need more than one.

 **Your Investigation**

Now it's your turn to do your investigation and write about it. Write about the following checklist items in your science notebook.

- [ ] **Choose a question or make up one of your own.**
- [ ] **Gather the materials you will use.**

Model

My Hypothesis

If I dip the coffee filter in water, then the ink from the marker will spread across the filter. I will be able to see the colors that are present in the black ink. If I test the ink of 2 different brands of black markers, then I will see the same colors in both types of ink.

You can use "If…, then…." statements to make your hypothesis clear.

 **Your Investigation**

☐ **If needed, make a hypothesis or prediction.**

Write your hypothesis or prediction in your science notebook.

Model

Variable I Will Change

I will change the brand of marker I use to draw on the coffee filter.

Variable I Will Observe or Measure

I will observe the colors that separate from the ink when it gets wet.

Variables I Will Keep the Same

Everything else will be the same. I will use the same type of filters and the same amount of water.

Answer these three questions:
1. What one thing will I change?
2. What will I observe or measure?
3. What things will I keep the same?

Your Investigation

☐ **If needed, identify, manipulate, and control variables.**

Write about the variables for your investigation.

Model

My Plan

1. Label the plastic cups **A** and **B**. Pour 50 mL of water in each cup.

2. Cut 2 strips from a coffee filter. Each strip should be about 3 cm by 7 cm.

3. Label 2 black markers **A** and **B**. Use marker A to draw a thick line about 1 cm from the bottom of one coffee filter strip.

4. Tape the strip to a craft stick and place the craft stick across the top of cup A. The bottom of the strip should just be touching the water in the cup.

5. Repeat steps 3 and 4 for cup B. Let the cups sit for 30 minutes.

6. Remove the strips and lay them on a flat surface to dry.

> Write detailed plans. Another student should be able to repeat your investigation without asking any questions.

SCIENCE ## Your Investigation

☐ **Make a plan for your investigation.**

Write the steps for your plan.

Model

I carried out all 6 steps of my plan.

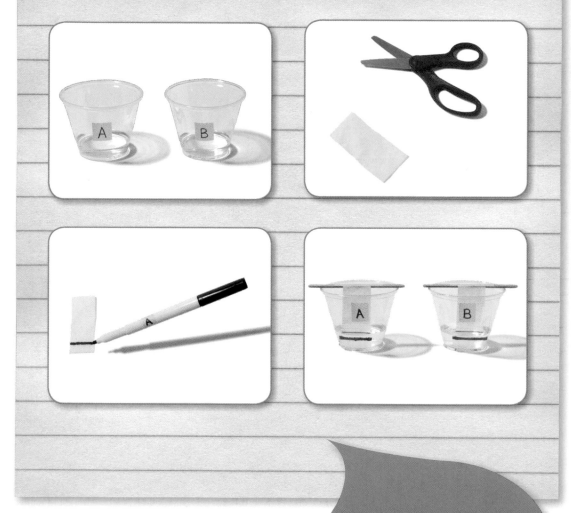

You might make a note if you needed to adjust your plan in any way. Cara had no changes to make.

 **Your Investigation**

☐ **Carry out your plan.**

Be sure to follow your plan carefully.

Model

Data (My Observations)

Colors Visible in Black Ink

	Start	30 Minutes	After Drying
Marker A	black	black, spread across strip	black, dark blue, reddish purple
Marker B	black	black, spread across strip	black

My Analysis

When I first drew on the coffee filter strips, the black ink from marker A and marker B looked the same. After soaking in the water for 30 minutes, the ink from both markers had spread out across the strip. After drying, I could see different colors in the ink from marker A, but the ink from marker B showed only one color.

Explain what happened based on the data you collected.

 **Your Investigation**

◻ **Collect and record data. Analyze your data.**

Collect and record your data, and then write your analysis.

Model

How I Shared My Results

I made a presentation to the class. First I showed what the black ink from the 2 markers looked like. Next I shared the results from the investigation. Then I told how my investigation shows the colors that are in black ink.

Scientists often share results so others can find out what was learned.

My Conclusion

All black ink is not alike. The colors in the black ink from the 2 markers were different. The results of the investigation show that my hypothesis was not supported.

Tell what you conclude and what evidence you have for your conclusion.

Another Question

I wonder what would happen if I tested ink from blue or green markers. Would the ink spread out to show more than one color?

Investigations often lead to new questions for Inquiry.

SCIENCE Your Investigation

☐ Explain and share your results.

☐ Tell what you conclude.

☐ Think of another question.

How Scientists Work

Using Data to Analyze Explanations

When scientists analyze the explanations of others, they look for facts. A fact is a statement that is backed up by a lot of data. It is a fact that when the temperature of liquid water at sea level rises to 100 °C, the liquid water begins to change to a gas called water vapor. Data has been collected many times to show that water at sea level boils when its temperature reaches 100 °C.

Scientists do not base their explanations on opinions. An opinion is a statement that is not based on evidence. It is what someone thinks about something. One person's opinion may be that the water in a pool, which has a temperature of 25 °C, feels warm. Another person might think that the water in the pool feels cool.

Collecting and Analyzing Data to Support an Explanation

Scientists avoid presenting opinions in scientific explanations. Instead, scientists collect evidence in the form of data during investigations. They analyze data and use the resulting information to support explanations.

A scientist hypothesized that if she dropped 2 objects with the same mass but different shapes, then 1 object would fall faster than the other. The scientist designed an experiment to gather data that would help her test this hypothesis.

The scientist used a marble and a piece of cardboard, each with the same mass. She dropped both objects from the same height. She measured with a stopwatch how long each object took to hit the floor. She repeated the test 5 times.

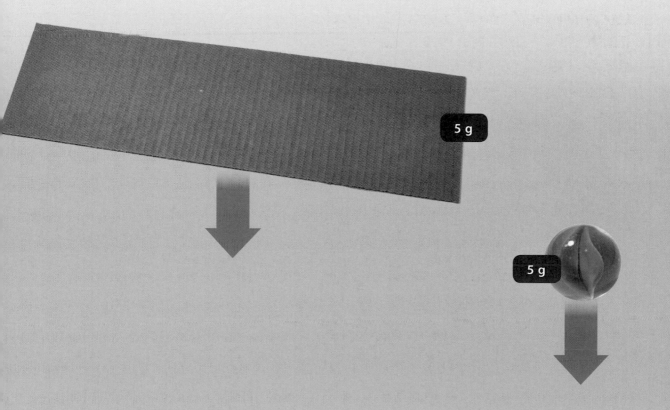

5 g

5 g

The scientist recorded her data in a table. Then the scientist analyzed her data and concluded that objects with the same mass but different shapes do fall at different speeds. The marble fell faster than the cardboard. Her data supported her conclusion. Others who look at her explanation can analyze her data to see whether the explanation is based on fact.

Marble

Trial	Time It Took to Fall 2 Meters (s)
1	1.52
2	1.37
3	1.43
4	1.32
5	1.50

Cardboard

Trial	Time It Took to Fall 2 Meters (s)
1	1.71
2	1.93
3	1.87
4	1.93
5	1.85

SUMMARIZE

What Did You Find Out?

1 How is a fact different from an opinion?

2 Why do scientists collect and analyze data during an investigation?

3 What do scientists look for when they analyze explanations?

Use Data to Support an Explanation

Carlos wants to find out if his toy car will move at the same speed on different surfaces. After observing the surfaces of three different materials, Carlos predicts that the car will move more slowly over rougher materials. He tests the car on three ramps with the same slope. He lets the car go at the top of the ramp and times how long it takes to reach the bottom of the ramp. He repeats the experiment four times and records his data in the chart below.

Material on the Ramp	Time (s)			
	Trial 1	Trial 2	Trial 3	Trial 4
Cardboard	1.62	1.60	1.65	1.66
Book	1.39	1.37	1.31	1.35
Sandpaper	2.37	2.40	2.25	2.35

1 Do Carlos's results support his prediction? Explain?

2 Analyze Carlos's results to state and support an explanation about how the car moved on different surfaces.

ACKNOWLEDGMENTS Grateful acknowledgment is given to the authors, artists, photographers, museums, publishers, and agents for permission to reprint copyrighted material. Every effort has been made to secure the appropriate permission. If any omissions have been made or if corrections are required, please contact the Publisher.

ILLUSTRATOR CREDITS 152 Precision Graphics. All maps by Mapping Specialists.

PHOTOGRAPHIC CREDITS Front, Back Cover Carsten Peter/National Geographic Image Collection. **1-2, 3-4, 5-6, 7-8, 9-10** Carsten Peter/National Geographic Image Collection. **13** Luis Marden/National Geographic Image Collection. **14** (tl) Ingram Publishing/Superstock. (tc) Stockbyte/Getty Images. (tr) C Squared Studios/Photodisc/Getty Images. (bl) Elena Butinova/Shutterstock. (br) Feng Yu/Shutterstock. **15** (t) Design Pics Inc./Alamy Images. (b) Birute Vijeikiene/Shutterstock. **16** (c) Image Source Pink/Jupiterimages. (bl) chudoba/Shutterstock. (br) Dario Sabljak/Shutterstock. **16-15** (t) Brandon Seidel/Shutterstock. **18** (t) Brandon Seidel/Shutterstock. (b) Courtesy of WARD's Natural Science Est.. **19** (b) Bryant Jayme/Shutterstock. **20** (t) Garry DeLong/Photo Researchers, Inc.. (cl, cr, b) Nigel Cattlin/Visuals Unlimited. **21** (bl) Biodisc/Visuals Unlimited. **22** (l) d Reschke/Peter Arnold, Inc.. (r) David Sieren/Visuals Unlimited. **23** (t) Garry DeLong/Photo Researchers, Inc.. (c) irabel8/Shutterstock. (b) April D/Shutterstock. **24** (t) PhotoDisc/Getty Images. (cl) BananaStock/Jupiterimages. (cr) Creatas/Jupiterimages. (b) John Foxx Images/Imagestate. **25** (t) PhotoDisc/Getty Images. **26** (t) PhotoDisc/Getty Images. **27** (b) Creatas/Jupiterimages. **28** (t) mashe/Shutterstock. (c) Vinicius Tupinamba/Shutterstock. **30** (t) mashe/Shutterstock. **31** (b) Don Paulson/Alamy Images. **32** (t) iStockphoto. (c) Image Source/Getty Images. (b) Paul Zahl/National Geographic Image Collection. **33** (t) Rick & Nora Bowers/Alamy. **34** (t) iStockphoto. **35** (b) Kim Taylor/Minden Pictures. **36** (t) DigitalStock/Corbis. **38** (t) DigitalStock/Corbis. **40** (t) Shaun Hensher/Shutterstock. (c) iStockphoto. **42** (t) Shaun Hensher/Shutterstock. **43** (b) PureStock/SuperStock. **44** (cl) Justaman/Shutterstock. (cr) DEA/C DANI/Photolibrary. **46** (ct) Noah Strycker/Shutterstock. (c) Amy McNabb/Shutterstock. (cb) Oshvintsev Alexander/Shutterstock. **47** (b) Eky Chan/Shutterstock. **48** (c) Konrad Wothe/Minden Pictures/National Geographic Image Collection. (b) Piotr Naskrecki/Minden Pictures/National Geographic Image Collection. **51** (b) Bates Littlehales/National Geographic Image Collection. **52** (c) Premier Edition Image Library/Superstock. (b) Miguel Angelo Silva/iStockphoto. **52-53** (t) kohy/Shutterstock. **54** (t) kohy/Shutterstock. **55** (b) Creatas/Jupiterimages. **56** (t) oriontrail/Shutterstock. (b) Creatas/Jupiterimages. **58** (t) oriontrail/Shutterstock. **59** (b) BrandX/Jupiterimages. **60** (t) Smit/Shutterstock. **63** (b) Andrew McClenaghan/Photo Researchers, Inc.. **64** (t) Laurence Monneret/Getty Images. (b) BananaStock/Jupiterimages. **66** (t) Laurence Monneret/Getty Images. **67** (b) Jacek Chabraszewski/Shutterstock. **68** (t) PhotoDisc/Getty Images. **69** Hans Pfletschinger/Science Faction/Corbis. **70** (b) Nature's Images/Photo Researchers, Inc.. **70-77** (t) Thomas Payne/Shutterstock. **78** (t) Andrew Paterson/Getty Images. (c) LWA/Sharie Kennedy/Blend Images/Corbis. (b) Darren Baker/Shutterstock. **79** Alexander Raths/Shutterstock. **80** (t) Nina Leen/Time & Life Pictures/Getty Images. (b) Grafissimo/iStockphoto. **81** (bg) Heidi & Hans-Jurgen Koch/Minden Pictures/National Geographic Image Collection. (t) Andrew Paterson/Getty Images. **86** (t) Visuals Unlimited/Corbis. (c) Roger Ressmeyer/Corbis. (bl) Ekaterina Starshaya/iStockphoto. (br) peresanz/Shutterstock. **88** (t) Visuals Unlimited/Corbis. **89** (b) Michael Freeman/Corbis. **90** (t) PhotoDisc/Getty Images. (b) Corbis Premium RF/Alamy Images. **92** (t) PhotoDisc/Getty Images. **93** (b) PhotoDisc/Getty Images. **94** (b) Forster Forest/Shutterstock. **94-95** (t) DSGpro/iStockphoto. **95** (c) Frans Lanting/National Geographic Image Collection. (b) Park Yeong-Dae/AFP/Getty Images. **96** (b) Flip Nicklin/Minden Pictures/National Geographic Image Collection. **96-97** (b) DSGpro/iStockphoto. **97** (b) Paul S. Howell/Getty Images. **98** (b) amana images inc./Alamy Images. **98-99** (t) Chernetskiy/Shutterstock. **100-101** (t) Chernetskiy/Shutterstock. **102-103** (t) John Foxx Images/Imagestate. **103** (cr) Denis Selivanov/Shutterstock. **104** (t) John Foxx Images/Imagestate. **105** (b) Stephen Alvarez/National Geographic Image Collection. **106-108** (t) Laitr Keiows/Shutterstock. **109** (b) Patrick Laverdant/iStockphoto. **110** (c) PhotoDisc/Getty Images. **110-111** (t) Stockbyte/Getty Images. **112** (t) Stockbyte/Getty Images. **113** (b) Bold Stock/Unlisted Images. **114-115** (t) PhotoDisc/Getty Images. **116** (t) PhotoDisc/Getty Images. **117** (b) Will Stanton/Alamy Images. **118-119, 120** (t) Photodisc/Getty Images. **121** (b) RiverNorthPhotography/iStockphoto. **122, 124** (t) Donna K. & Gilbert M. Grosvenor/National Geographic Image Collection. **125** (b) James P. Blair/National Geographic Image Collection. **126-127** (t) DigitalStock/Corbis. **128** (t) DigitalStock/Corbis. **129** (b) Karen Kasmauski/National Geographic Image Collection. **130-131** (t) Kenneth Sponsler/Shutterstock. **132** (t) Kenneth Sponsler/Shutterstock. **133** (b) Michael McCloskey/Photodisc/Getty Images. **134** (b) Michaela Rehle/Reuters/Corbis. **134-135, 136** (t) John Foxx Images/Imagestate. **137** (b) Bruce Heinemann/PhotoDisc/Getty Images. **138** (c) Joseph Scott Photography/Shutterstock. (b) James P. Blair/National Geographic Image Collection. **138-139, 140** (t) PhotoDisc/Getty Images. **141** (b) Bruce Heinemann/Getty Images. **142-143** (t) John Foxx Images/Imagestate. **143** (b) Carsten Peter/Speleoresearch & Films/National Geographic Image Collection.

144 (b) sciencephotos/Alamy Images. **144-151** (t) John Foxx Images/Imagestate. **152** (b) Alaska Stock Images/National Geographic Image Collection. **152-153** (t) Josh Westrich/Corbis. **153** Teuton, T. C., Main, C. L., Mueller, T. C., Wilkerson, J. B., Brecke, B. J., Unruh, J. B. 2005. Prediction modeling for tropical signalgrass (Urochloa subquadripara) emergence in Florida. Online. Applied Turfgrass Science doi:10.1094/ATS-2005-0425-01-BR.. **154-155** (t) Josh Westrich/Corbis. **157** (b) Georgy Markov/iStockphoto. **160, 162** (t) FotografiaBasica/iStockphoto. **163** (b) scoutingstock/Shutterstock. **164** (t) PhotoDisc/Getty Images. (b) Steve Cole/iStockphoto. **167** (t) PhotoDisc/Getty Images. **168** (t) John McLaird/Shutterstock. (c) Steve Hix/Somos Images/Corbis. (b) Gary Paul Lewis/Shutterstock. **170** (t) John McLaird/Shutterstock. **171** (b) SF photo/Shutterstock. **172** (t) Fancy/Veer/Corbis. (c) Mikael Damkier/iStockphoto. (b) Robert Adrian Hillman/Shutterstock. **174** (t) Fancy/Veer/Corbis. **175** (b) Creatas/Jupiterimages. **176, 178** (t) Artem Tovstinchuk/iStockphoto. **179** (b) Jam.si/Shutterstock. **180** (t) Only Fabrizio/Shutterstock. (c) The Dr. Samuel D. Harris National Museum of Dentistry. (b) Chas/Shutterstock. **181** (t, b) The Dr. Samuel D. Harris National Museum of Dentistry. (c) SSPL/Getty Images. **182** Philippe Psaila/Photo Researchers, Inc.. **183** (t) Only Fabrizio/Shutterstock. (c, b) Philippe Psaila/Photo Researchers, Inc.. **184** Philippe Psaila/Photo Researchers, Inc.. **185** (t) Only Fabrizio/Shutterstock. **186** (t) Martin Diebel/Getty Images. (b) Mike Kemp/Rubberball Productions/Getty Images. **188** (t) Martin Diebel/Getty Images. **189** (b) James Steidl/Shutterstock. **190** (t) Mark Scheuern/Alamy Images. **192** (t) Mark Scheuern/Alamy Images. **193** (b) Lykovata/Shutterstock. **194** (t) iStockphoto. (c) Richard Megna/Fundamental Photographs, NYC. **196** (t) iStockphoto. **197** (b) Stephen Derr/Photographer's Choice/Getty Images. **198** (t) Awe Inspiring Images/Shutterstock. **200** (t) Awe Inspiring Images/Shutterstock. **202** (t) Jaimie Duplass/Shutterstock. (c) Taylor Hinton/iStockphoto. (b) iStockphoto. **204** (t) Jaimie Duplass/Shutterstock. **205** (b) Image Source. **206-208** (t) Michael W. Davidson/Photo Researchers, Inc.. **209** (b) sjeacle/Shutterstock. **210** (t) Ekaterina Starshaya/Shutterstock. (c) PhotoDisc/Getty Images. (b) Yevgen Timashov/Shutterstock. **211, 212** (t) Ekaterina Starshaya/Shutterstock. **213** (b) Stockbyte/Getty Images. **214** (b) Tetra Images/Jupiterimages. **214-215, 216** (t) Stockbyte/Getty Images. **217** (b) Artville. **218** (c) Natalya Ilinskaya/iStockphoto. (bl) koksharov dmitry/iStockphoto. (br) nikkytok/Shutterstock. **218-219, 220** (t) Petr Jilek/Shutterstock. **221** (b) Frederic Pitchal/Sygma/Corbis. **222-223, 224** (t) Max Bukovski/Shutterstock. **225** (b) Alan and Sandy Carey/Jupiterimages. **226, 227** (t) Geoff Tompkinson/Photo Researchers, Inc.. **228, 230-233, 235** (t) Edmond Van Hoorick/PhotoDisc/Getty Images. **236** (t) PhotoDisc/Getty Images. (b) Roy Morsch/Corbis. **237** (t) Eyewire/Getty Images. **239** (t) PhotoDisc/Getty Images.

PROGRAM AUTHORS Judith Sweeney Lederman, Ph.D., Director of Teacher Education and Associate Professor of Science Education, Department of Mathematics and Science Education, Illinois Institute of Technology, Chicago, Illinois; Randy Bell, Ph.D., Associate Professor of Science Education, University of Virginia, Charlottesville, Virginia; Malcolm B. Butler, Ph.D., Associate Professor of Science Education, University of South Florida, St. Petersburg, Florida; Kathy Cabe Trundle, Ph.D., Associate Professor of Early Childhood Science Education, The Ohio State University, Columbus, Ohio; David W. Moore, Ph.D., Professor of Education, College of Teacher Education and Leadership, Arizona State University, Tempe, Arizona

THE NATIONAL GEOGRAPHIC SOCIETY
John M. Fahey, Jr., President & Chief Executive Officer
Gilbert M. Grosvenor, Chairman of the Board

National Geographic School Publishing
Hampton-Brown
www.NGSP.com

Printed in the USA. RR Donnelley, Menasha, WI

ISBN: 978-0-7362-7779-2

12 13 14 15 16 17 18 19 20

6 7 8 9 10